T0320720

Edited by Editorial Board of UTokyo Engineering Course

Partial
Differential
Equations

UTokyo Engineering Course/Basic Mathematics

Linear Algebra I: Basic Concepts
by Kazuo Murota and Masaaki Sugihara
ISBN: 978-981-125-702-5
ISBN: 978-981-125-797-1 (pbk)

Linear Algebra II: Advanced Topics for Applications
by Kazuo Murota and Masaaki Sugihara
ISBN: 978-981-125-705-6
ISBN: 978-981-125-798-8 (pbk)

Complex Function Theory
by Takeo Fujiwara
ISBN: 978-981-127-091-8
ISBN: 978-981-127-132-8 (pbk)

UTokyo Engineering Course / Basic Mathematics

Edited by Editorial Board of UTokyo Engineering Course

Partial
Differential
Equations

Osamu Sano
Professor Emeritus at Tokyo University of
Agriculture and Technology, Japan

Published by

World Scientific Publishing Co. Pte. Ltd.
5 Toh Tuck Link, Singapore 596224
USA office: 27 Warren Street, Suite 401-402, Hackensack, NJ 07601
UK office: 57 Shelton Street, Covent Garden, London WC2H 9HE

and

Maruzen Publishing Co., Ltd.
Kanda Jimbo-cho Bldg. 6F, Kanda Jimbo-cho 2-17
Chiyoda-ku, Tokyo 101-0051, Japan

British Library Cataloguing-in-Publication Data
A catalogue record for this book is available from the British Library.

UTokyo Engineering Course/Basic Mathematics
PARTIAL DIFFERENTIAL EQUATIONS

Copyright © 2023 by Osamu Sano

ISBN 978-981-127-088-8 (hardcover)
ISBN 978-981-127-131-1 (paperback)
ISBN 978-981-127-089-5 (ebook for institutions)
ISBN 978-981-127-090-1 (ebook for individuals)

For any available supplementary material, please visit
https://www.worldscientific.com/worldscibooks/10.1142/13266#t=suppl

Desk Editor: Tan Rok Ting

Printed in Singapore

UTokyo Engineering Course

About This Compilation

What is the purpose of engineering education at the University of Tokyo's Undergraduate and Graduate School of Engineering? This School was established 125 years ago, therefore we feel it is an appropriate time to ask this question again. More than a century has passed since Japan embarked on a path to introduce and negotiate Western knowledge and practices. Japan and the world are very different places now, and today our university stands as a world leading institute in engineering research and education. As such, it is our duty and mission to build a firm foundation of education that will support the creation and dissemination of engineering knowledge, practices and resources. Our School of Engineering must not only teach outstanding students from Japan but also those from throughout the world. Put another way, the engineering that we teach students is not only a responsibility of this School, but an imperative placed on us by society and the age in which we live. It is in this changed context, where we have gone from follower to leader, that we present this curriculum, The University of Tokyo (UTokyo) Engineering Course. The course is a reflection of the School's desire to engage with those outside the walls of the Ivory Tower, and to spread the best of engineering knowledge to the world outside our institution. At the same time, the course is also designed for the undergraduate and graduate students of the School. As such, the course contains the knowledge that should be learnt by our students, taught by our instructors and critically explored by all.

February 2012

Takehiko Kitamori

Dean, Undergraduate and Graduate Schools of Engineering

The University of Tokyo

(April 2010–March 2012)

UTokyo Engineering Course

The Purpose of This Publication

Modern engineering is composed of the academic discipline of fundamental engineering and the academic discipline of integrated engineering that deals with specific systems and subjects. Interdisciplinary disciplines and multidisciplinary disciplines are amalgamations of multiple academic disciplines that result in new academic disciplines when the academic pursuit in question does not fit within one traditional fundamental discipline. Such interdisciplinary disciplines and multidisciplinary disciplines, once established, often develop into integrated engineering. Moreover, the movement toward interdisciplinarity and multidisciplinarity is well underway within both fundamental engineering and advanced research.

These circumstances are producing a variety of challenges in engineering. That is, the scope of research of integrated engineering is gradually growing larger, with economics, medicine and society converging into an enormously complex social system, which is resulting in the trend of connotative academic disciplines growing larger and becoming self-contained research fields, which, in turn, is resulting in a trend of neglect toward fundamental engineering. The challenge of fundamental engineering is how to connect engineering education that is built upon traditional disciplines with that of advanced engineering research in which interdisciplinarity and multidisciplinarity is continuing at a rapid pace. Truly this is an educational challenge shared by all the top engineering schools in the world. Without having a solid understanding of engineering, however, education related to learning state-of-the-art research methodologies will not hold up. This is the dichotomy of higher education in engineering; that is, higher education in engineering simply will not work out if either side of the equation is missing.

In the meantime, the internationalization of universities is going forward in routine fashion. In fact, here at the University of Tokyo (UTokyo), one quarter of the graduate students enrolled in engineering fields are of foreign nationality and the percentage of foreign undergraduate students is expected to increase more and more. On top of that, Japan is experiencing a reduction in the population of its youth. Therefore, the time is ripe to ramp up efforts to look outside of Japan in order to secure the human resources to sustain the future of advanced science and technology here in Japan. It is clear that the internationalization of engineering education is rapidly underway. As such, the need for a curriculum that is firmly rooted in engineering knowledge needs to be oriented toward both local and foreign students.

Due to these circumstances surrounding modern engineering, we at UTokyo's School of Engineering have systematically organized an engineering curriculum of fundamental engineering knowledge that will not be unduly influenced by the times, with the goal of firmly establishing a benchmark suitable for that of the top schools of engineering of science and technology for students to learn and teachers to teach. This engineering curriculum clarifies the disciplines and instruction policy of UTokyo's School of Engineering and is composed of three layers: Fundamental (sophomores (second semester) and juniors), Intermediate (seniors and graduates) and Advanced (graduates). Therefore, this engineering course is a policy for the thorough education of the engineering knowledge necessary for forming the foundation of our doctorate program as well. The following is an outline of the expected effect of this engineering course:

- Surveying the total outline of this engineering course will assist students in understanding which studies they should undertake for each field they are pursuing, and provide an overall image by which the students will know what fundamentals they should be studying in relation to their field.
- This course will build the foundation of education at UTokyo's School of Engineering and clarify the standard for what instructors should be teaching and what students need to know.
- As students progress in their major it may be necessary for them to go back and study a new fundamental course. Therefore the textbooks are designed with such considerations as well.

- By incorporating explanations from the viewpoint of engineering departments, the courses will make it possible for students to learn the fundamentals with a constant awareness of their application to engineering.

Yasuhiro Kato, Board Chair
Yukitoshi Motome, Shinobu Yoshimura, Executive Secretary
Editorial Board of UTokyo Engineering Course

xi

Preface

A partial differential equation (hereafter abbreviated as PDE) describes the space- and/or time-dependent phenomena in a wide range of science and engineering, where a number of basic problems, not only in mathematics, but also practical problems in industry, agriculture, and socio-economics among other areas given by PDEs, are awaiting solutions. Because of the expanding application fields with increased complexity and different types of behaviors to be analyzed, it is not easy to summarize in a word what a PDE is. However, if we extract the essential principles underlying the physical phenomena depending on two or more independent variables, thereby discarding the detailed differences particular to the respective problems, they will fall into common features inherent in PDEs.

In Chapter 1, we introduce a first-order PDE. With a PDE of a lower order differentiation and the smallest possible number of variables, we try to construct an intuitive understanding that connects mathematical expressions with the geometric interpretations. This chapter serves as a first step toward the PDEs of a higher-order and/or toward those depending on many more variables.

In Chapter 2, we deal with second-order PDEs, which are particularly important in science and engineering. We classify them into three types: hyperbolic, parabolic and elliptic; or in other words, wave equation, diffusion equation (or heat conduction equation) and Laplace–Poisson equation, respectively, reflecting their physical backgrounds. We show typical methods of solution, such as a separation of variables that reduces the PDE to a simpler set of equations, or eigenfunction expansions that play roles of a kind of "unit vectors" in a functional space. For practical problems, we need to obtain the solutions that satisfy the boundary conditions. For such purposes, several coordinate systems have been developed. The coordinate systems whose coordinate curves or surfaces coincide with the boundary shape are appropriate for analyzing the given PDE. As a basic tool, we show typical coordinate systems that are often used in practice. To obtain the full solutions, we need to know both regular solutions within the region under consideration and the principal solutions that reflect the singularity. The latter is given by the Green's function, which connects the so to speak an "external force" with its "response". Readers will learn the essential points on these solution methods, along with relevant examples.

In Chapter 3, we introduce a method of solution using integral transforms, such as Fourier, Laplace, Mellin and Hankel transforms. The choice of transforms depends on the spatial geometry and the types of boundary and/or initial conditions. Once an integral transform is applied to a PDE, the number of independent variables in the original differential equation is reduced, so that the solution in the transformed space becomes easier to obtain. By applying the inverse transforms, the solution in the original space is attained.

The present textbook was originally written as part of a series of the Mathematics/UTokyo Engineering Course published in Japanese, so that the fundamental knowledge of calculus, complex function theory, ordinary differential equation, vector analysis, among others, are assumed to be known. Although this textbook aims at presenting the basic knowledge of PDEs as self-contained as possible, some readers may feel the description of particular mathematics to be too short, or not accurate enough. In the present textbook, however, emphasis is not put on rigorous mathematics, but put on providing methods to find the solution of the given problems. Readers who feel their understanding in particular areas inadequate are recommended to consult relevant reference books.

For scientists and engineers working in today's highly specialized and deeply interconnected circumstances, needs to overcome the confronted problems beyond the existing knowledge are increasing. Under such situations, logical thinking based on systematic knowledge in one branch of science and engineering, and an innovative spirit to apply the prevailing knowledge to other branches, will help constructing a network of knowledges from the viewpoint of PDEs, which will provide the readers with an insight into new approaches and help them find clues to reach the goals.

O. Sano

Contents

Preface   xiii

1. First-order Partial Differential Equations   1

    1.1   Linear PDEs 1

    1.2   First-order PDEs and Characteristic Strips 7

        1.2.1   Integral surfaces 7

        1.2.2   Characteristic curves and characteristic strips . . 8

        1.2.3   Solution of the first-order quasi-linear PDEs . . . 18

    1.3   Initial-value Problems of the First-order PDEs 21

    1.4   Complete Integrals 23

        1.4.1   Complete solutions, general solutions, and singular

              solutions 23

        1.4.2   First-order PDEs with easily obtainable complete

              solutions 27

        1.4.3   Integrable condition 29

        1.4.4   Lagrange–Charpit method 31

    1.5   Hamilton–Jacobi Theory 36

        1.5.1   Extremum problems and the Euler–Lagrange

              equation 36

        1.5.2   Lagrange's equation and canonical equation 42

        1.5.3   Canonical transform 43

        1.5.4   Hamilton–Jacobi equation 45

    1.6   Integral of the Hamilton–Jacobi equation 46

2. Second-order Partial Differential Equations   51

    2.1   Examples of Second-order PDEs 51

2.1.1   Wave equation   51
2.1.2   Diffusion equation   53
2.1.3   Laplace–Poisson equation   54
2.2   Classification of Second-order PDEs   55
2.2.1   Classification (hyperbolic, elliptic, parabolic) . . .   55
2.2.2   Integral surfaces and initial-value problems   59
2.3   Separation of Variables and Eigenvalue Problems   65
2.3.1   Orthogonal curvilinear coordinates and separation
        of variables   66
2.3.2   Eigenvalues and eigenfunctions in the
        Sturm–Liouville equations   80
2.3.3   Eigenfunction expansions   83
2.3.4   Fourier–Bessel Expansions   91
2.4   Green's Function and the Boundary-value Problems . . .   93
2.4.1   Delta function   93
2.4.2   Adjoint PDE and Green's formulae   95
2.4.3   Green's function   97
2.4.4   Principal solution of the Laplace equation   98
2.4.5   Adjoint Green's function and reciprocity principle   105
2.4.6   Principal solution of the wave equation   106
2.4.7   Principal solution of the diffusion equation   110
2.5   Initial- and/or Boundary-value Problems in Rectangular
      Regions   113
2.5.1   Laplace equation   113
2.5.2   Wave equation   118
2.5.3   Diffusion equation   122
2.6   Initial- and/or Boundary-value Problems in Circular or
      Spherical Regions   127
2.6.1   Dirichlet's problems for circular or spherical regions   127
2.6.2   Vibration of a circular membrane   133
2.6.3   Diffusion in a circular region   136
2.6.4   Radiation and scattering of waves   136

3. Integral Transforms and their Applications   141

3.1   General Theory of Integral Transforms and their
      Applicability   141
3.2   Integral Transforms in a Finite Region   145
3.2.1   General theory   145
3.2.2   Further applications   147

3.3   Integral Transforms in an Infinite Region 148
      3.3.1   General theory 148
      3.3.2   Fourier transforms 152
      3.3.3   Laplace transforms 167
      3.3.4   Mellin transforms 176
      3.3.5   Hankel transforms 187
      3.3.6   Application of integral transforms to integral
              equations 194

Appendix A   201

A.1   Gamma Function and Beta Function 201
      A.1.1   Gamma function 201
      A.1.2   Beta function 202
A.2   Bessel Functions 202
      A.2.1   Bessel functions and Neumann functions 202
      A.2.2   Hankel functions 207
      A.2.3   Spherical Bessel functions 207
      A.2.4   Modified Bessel functions 208
A.3   Legendre Functions 209
      A.3.1   Legendre functions 210
      A.3.2   Associated Legendre functions 212
      A.3.3   Spherical surface harmonic functions 213

Bibliography   215

Index   217

Chapter 1

First-order Partial Differential Equations

A partial differential equation (hereafter referred to "PDE") is a relation between two or more independent variables $x, y, \ldots$ and a dependent variable u, as well as the partial derivatives of the latter with respect to the former variables, $i.e.$,

$$f(x, y, \ldots, u, u_x, u_y, \ldots, u_{xx}, u_{xy}, \ldots) = 0, \qquad (1.1)$$

where the dependent variable u is the unknown function to be determined. Subscripts of u denote the partial derivatives of u with respect to the relevant variables, such as

$$u_x = \frac{\partial u}{\partial x}, \quad u_y = \frac{\partial u}{\partial y}, \quad u_{xx} = \frac{\partial^2 u}{\partial x^2}, \quad u_{xy} = \frac{\partial^2 u}{\partial x \partial y}.$$

In this chapter, we deal with first-order PDEs. We first examine the fundamental properties underlying the simplest PDEs that depend on two independent variables, and describe their geometrical interpretation. We then extend the basic knowledge to general PDEs involving many more independent variables.

1.1 Linear PDEs

Consider, for example, a circle given by

$$x^2 + y^2 = a^2, \qquad (1.2)$$

where (x, y) is the Cartesian coordinate system on the plane of the circle. In the above expression, a constant a that specifies a particular radius appears. To show the *general* feature of a circle, however, such a constant is not necessary. For the latter purpose, we need to eliminate the constant a from Eq. (1.2), and show only the relation between x, y, and the derivatives with respect to them. Since x and y are equivalent in the expression (1.2), we may take one of the variables x (or y) as an independent variable, then the other variable y (or x) becomes a dependent variable. Choosing the former, for instance, yields

$$\frac{dy}{dx} = -\frac{x}{y}, \qquad (1.3)$$

by differentiating Eq. (1.2) with respect to x.[1] The equation (1.3), which is an ordinary differential equation (hereafter referred to "ODE") in this case, provides the characteristics of a circle.

The above example refers to a curve in a two-dimensional space. How are the surfaces in a three-dimensional space described? To see this, we introduce a three-dimensional Cartesian coordinate system (x, y, z). We first consider a surface of revolution with the z axis as an axis of rotation, so that the cross section in the plane parallel to the xy plane is a circle whose radius changes with height z.

Example 1.1 (Spheroid). A spheroid, with the z axis as an axis of rotation, is given by

$$\frac{x^2 + y^2}{a^2} + \frac{z^2}{c^2} = 1, \qquad (1.4)$$

where a is the radius of the circle in the xy plane, and c is the semi-major (or semi-minor) axis of the ellipse in the z direction. We eliminate a and c to show the general feature of a spheroid. We first regard x and y as independent variables, whereas z as a dependent variable, and carry out a partial differentiation of Eq. (1.4) with respect to x and y. We then have

$$\frac{x}{a^2} + \frac{z}{c^2} p = 0, \qquad \frac{y}{a^2} + \frac{z}{c^2} q = 0, \qquad (1.5)$$

where the customary notations

$$p \equiv \frac{\partial z}{\partial x}, \qquad q \equiv \frac{\partial z}{\partial y} \qquad (1.6)$$

---

[1]The above relation may be obtained by differentiating Eq. (1.2) term by term with respect to x,

$$2x + 2y\frac{dy}{dx} = 0,$$

from which Eq. (1.3) follows. Alternatively, Eq. (1.3) may be obtained by solving Eq. (1.2) in the form $y = \pm\sqrt{a^2 - x^2}$, differentiating the latter with respect to x:

$$\frac{dy}{dx} = \pm\frac{-2x}{2\sqrt{a^2 - x^2}},$$

for which $y = \pm\sqrt{a^2 - x^2}$ is substituted to eliminate a, where the plus and minus signs ($\pm$) are accordingly maintained. Either procedure yields Eq. (1.3).

have been used. By eliminating a and c from Eq. (1.5), we have

$$yp - xq = 0, \tag{1.7}$$

which is the equation that describes the surface of a spheroid in general. ◁

• The number of operations in which a partial differentiation is carried out on the given dependent variable is called a **rank** or **order**. When the highest rank of the derivatives included in the given PDE is n, the latter equation is called an nth-**order** PDE. For example, Eq. (1.7) is a first-order PDE, since it includes utmost first-order partial derivative $\partial z/\partial x$ or $\partial z/\partial y$. In general, a first-order PDE of a given function $u(x, y)$ of two independent variables x, y is expressed in the form

$$f(x, y, u, p, q) = 0, \quad \text{where} \quad p = \frac{\partial u}{\partial x}, \quad q = \frac{\partial u}{\partial y}. \tag{1.8}$$

When Eq. (1.8) is expressed as a linear combination of the dependent variable u and its partial derivatives included (here, p and q), the equation is called a **linear** PDE. We provide a further classification at the end of this subsection.

Example 1.2. The PDE for a function $u(x, y)$ of two independent variables x and y, *e.g.*, Eq. (1.7) mentioned above, or

$$\frac{\partial u}{\partial x} = 0, \quad \text{or} \quad y\frac{\partial u}{\partial x} + x^2 = 0, \quad \text{etc.}$$

are all first-order linear PDEs. Furthermore,

$$\frac{\partial^2 u}{\partial x^2} = 0, \quad \text{or} \quad \frac{\partial^2 u}{\partial x \partial y} = 0, \quad \text{or} \quad \frac{\partial^2 u}{\partial x^2} + \frac{\partial^2 u}{\partial y^2} + \frac{\partial u}{\partial x} = 0, \quad \text{etc.}$$

are all second-order linear PDEs. ◁

Example 1.3 (Surfaces of revolution). A surface of revolution with the z axis as an axis of rotation is characterized by a circle of radius $r = \sqrt{x^2 + y^2}$ in the plane parallel to the xy plane, so that the surfaces of revolution are generally given by an equation of the form[2] $r = F_0(z)$, or conversely,

$$z = F_0^{-1}(r) \equiv F(r), \quad r = \sqrt{x^2 + y^2}, \tag{1.10}$$

---

[2]**(Functional relation)** Note that we may choose the surface of revolution using any function of $x^2 + y^2$. In general, if u is a function of g, and g is a function of $v(x, y)$, *i.e.*, $u(x, y) = F(v(x, y))$, then

$$p = \frac{\partial u}{\partial x} = \frac{\partial F}{\partial v}\frac{\partial v}{\partial x}, \quad q = \frac{\partial u}{\partial y} = \frac{\partial F}{\partial v}\frac{\partial v}{\partial y},$$

where r is the radius of the cross-sectional circle at position z. Here, F_0 (and its inverse function F) is an arbitrary function that depends on the problem, which is assumed to be differentiable up to the first order. To eliminate an arbitrary function F, we differentiate Eq. (1.10) with respect to x and y:

$$p \equiv \frac{\partial z}{\partial x} = F'(r)\frac{x}{r}, \qquad \text{and} \qquad q \equiv \frac{\partial z}{\partial y} = F'(r)\frac{y}{r},$$

from which we have

$$yp - xq = 0. \tag{1.11}$$

This is the PDE that is satisfied by the surfaces of revolution with arbitrary shape in the meridian plane, which is the same as Eq. (1.7). ◁

When Eq. (1.11) [or Eq. (1.7)] is solved, and the arbitrary function included in it is specified so as to satisfy the (boundary) condition at $y = 0$:

$$z^2 = F(x)^2 = c^2\left(1 - \frac{x^2}{a^2}\right),$$

it gives the ellipsoid of revolution given by Eq. (1.4). In contrast, if the boundary condition at $y = 0$ is

$$z^2 = F(x)^2 = c^2\left(\frac{x^2}{a^2} - 1\right),$$

then the solution gives a hyperboloid of revolution

$$\frac{x^2 + y^2}{a^2} - \frac{z^2}{c^2} = 1. \tag{1.12}$$

• As can be seen from these examples, an arbitrary function appears when the PDE is solved (integrated). In solving the ordinary differential equations (ODEs), we learned that a solution containing the same number

---

so that u satisfies the following PDE:

$$p\frac{\partial v}{\partial y} = q\frac{\partial v}{\partial x}, \quad \text{or} \quad \frac{\partial u}{\partial x}\frac{\partial v}{\partial y} - \frac{\partial u}{\partial y}\frac{\partial v}{\partial x} = 0.$$

Here, the expression

$$J \equiv \frac{\partial(u, v)}{\partial(x, y)} \equiv \frac{\partial u}{\partial x}\frac{\partial v}{\partial y} - \frac{\partial u}{\partial y}\frac{\partial v}{\partial x} \tag{1.9}$$

is called the **Jacobian**. Conversely, if $J \equiv \partial(u, v)/\partial(x, y) = 0$, it implies that u and v are not independent but are related as $u(x, y) = F(v(x, y))$, where F is an arbitrary function. In the present example, we reach the same equation, by any choice of v such as $v = x^2 + y^2$, or $v = \sqrt{x^2 + y^2}$, or $\cdots$, as long as the dependence on x and y of the form $x^2 + y^2$ is included.

of arbitrary constants as its order is called the general solution whereas a solution that satisfies particular conditions is called the particular solution. Similar terminologies are used for PDEs. Namely, when the solution of an nth-order PDE includes the same number of arbitrary functions as its order, it is called a **general solution** (or **general integral**). A particular solution that satisfies the given conditions is called a **particular solution** (or **particular integral**). In the above example, Eq. (1.10) is a general solution, whereas Eqs. (1.4) and (1.12) are particular solutions. See §§1.4.1 for further details on the types of solution.

Example 1.4. Among the PDEs for the function $u(x, y)$ of two independent variables x and y, as shown in Example 1.2, *e.g.*,
 - a general solution of the first-order PDE:

$$\frac{\partial u}{\partial x} = 0 \quad \text{is} \quad u(x, y) = \varphi(y), \quad (i.e., \text{ independent of } x),$$

 - a general solution of the second-order PDE:

$$\frac{\partial^2 u}{\partial x \partial y} = 0, \quad \text{is} \quad u(x, y) = \varphi(y) + \psi(x),$$

where, $\varphi(y)$ and $\psi(x)$ are arbitrary functions of y and x, respectively. (Confirm by substitution that $u(x, y)$ satisfies the respective equations.) ◁

If we take the limit $a \to 0$ under the condition $a/c \equiv k$ (=constant) in the hyperboloid of revolution (1.12), we have

$$x^2 + y^2 - \frac{a^2 z^2}{c^2} = a^2 \quad \to \quad x^2 + y^2 - k^2 z^2 = 0, \text{ or } \left(\frac{x}{z}\right)^2 + \left(\frac{y}{z}\right)^2 = k^2.$$

The latter gives a circular cone with the z axis as an axis of rotation. Then, how can a more general conical surface with an arbitrary cross section be described?

We know that a straight line passing through the origin in a three-dimensional xyz space is expressed by

$$z = C_1 x, \qquad z = C_2 y, \tag{1.13}$$

where C_1 and C_2 are arbitrary constants. By assigning appropriate values to the constants C_1 and C_2, the straight line of the desired orientation is specified. If the contour shape of a cone in a transverse section is given, the above-mentioned straight lines must follow the edge of that cross section. This condition imposes a certain functional relation between the constants C_1 and C_2:

$$F(C_1, C_2) = 0.$$

By eliminating C_1 and C_2 using $C_1 = z/x$ and $C_2 = z/y$, we obtain an equation that expresses the general conical surface:[3]

$$F\left(\frac{z}{x}, \frac{z}{y}\right) = 0. \tag{1.14}$$

Example 1.5 (Conical surfaces). We shall now find a PDE that is satisfied by the conical surface (1.14). For brevity, we introduce new variables $z/x = \xi$ and $z/y = \eta$, and perform partial differentiations on Eq. (1.14) with respect to the independent variables x and y. Then we have

$$\frac{\partial F}{\partial x} = \frac{\partial F}{\partial \xi}\frac{\partial \xi}{\partial x} + \frac{\partial F}{\partial \eta}\frac{\partial \eta}{\partial x} = \frac{\partial F}{\partial \xi}\left(\frac{p}{x} - \frac{z}{x^2}\right) + \frac{\partial F}{\partial \eta}\frac{p}{y} = 0, \quad \left(p \equiv \frac{\partial z}{\partial x}\right),$$

$$\frac{\partial F}{\partial y} = \frac{\partial F}{\partial \xi}\frac{\partial \xi}{\partial y} + \frac{\partial F}{\partial \eta}\frac{\partial \eta}{\partial y} = \frac{\partial F}{\partial \xi}\frac{q}{x} + \frac{\partial F}{\partial \eta}\left(\frac{q}{y} - \frac{z}{y^2}\right) = 0, \quad \left(q \equiv \frac{\partial z}{\partial y}\right).$$

The elimination of $\partial F/\partial \xi$ and $\partial F/\partial \eta$ from the above equations yields

$$z = xp + yq, \tag{1.15}$$

which is a PDE satisfied by the conical surfaces of an arbitrary cross-sectional shape.   ◁

Example 1.6 (PDE for $F(f(x, y, u), g(x, y, u)) = 0$). For the given functions $f(x, y, u)$ and $g(x, y, u)$, the intersection of the two surfaces

$$f(x, y, u) = C_1, \qquad g(x, y, u) = C_2, \quad (C_1, C_2 : \text{constants}),$$

gives a curve in the xyu space. If we impose a certain relation $F(C_1, C_2) = 0$ between the constants C_1 and C_2, a family of curves (a surface in this space) is formed. We shall find the PDE for which the latter surface $F(f, g) = 0$ satisfies. Here and hereafter we take it for granted that the first-order partial derivatives of F with respect to f and g exist unless otherwise stated. To remove F, we calculate the partial derivatives of F with respect to x and y:

$$\frac{\partial F}{\partial f}\left(\frac{\partial f}{\partial x} + \frac{\partial f}{\partial u}p\right) + \frac{\partial F}{\partial g}\left(\frac{\partial g}{\partial x} + \frac{\partial g}{\partial u}p\right) = 0,$$

---

[3]If the relation:

$$F(C_1, C_2) \equiv \frac{1}{C_1{}^2} + \frac{1}{C_2{}^2} - k^2 = 0$$

is imposed on F, then Eq. (1.14) gives a circular cone.

$$\frac{\partial F}{\partial f}\left(\frac{\partial f}{\partial y} + \frac{\partial f}{\partial u}q\right) + \frac{\partial F}{\partial g}\left(\frac{\partial g}{\partial y} + \frac{\partial g}{\partial u}q\right) = 0.$$

On eliminating $\partial F/\partial f$ and $\partial F/\partial g$ from the above equations, we obtain

$$p\,\frac{\partial(f,g)}{\partial(y,u)} + q\,\frac{\partial(f,g)}{\partial(u,x)} = \frac{\partial(f,g)}{\partial(x,y)}, \qquad (1.16)$$

where $\partial(f,g)/\partial(x,y)$ is the Jacobian defined by Eq. (1.9). Equation (1.16) is the PDE that is satisfied by F. ◁

- PDE (1.16) is formally written as

$$P(x,y,u)\,p + Q(x,y,u)\,q = R(x,y,u), \qquad (1.17)$$

which is regarded as a linear combination of the relevant partial derivatives (p and q in the above expression). Functions P, Q, and R are given functions of x, y, u, where P and Q are not assumed to be zero simultaneously ($P^2 + Q^2 \neq 0$). The type of PDE given above is called a **quasi-linear** PDE.[4] Equation (1.17) is also called the **Lagrange's partial differential equation**. When a PDE is described as a linear combination of the dependent variable (here, u) as well as the partial derivatives of the latter (here, p and q):

$$P(x,y)\,p + Q(x,y)\,q + R(x,y)\,u = S(x,y), \qquad (1.18)$$

it is called a **linear** PDE (P, Q, R, S are given functions of x and y, where $P^2 + Q^2 \neq 0$ is assumed).

Equation (1.7) of Example 1.1, Eq. (1.11) of Example 1.3, and Eq. (1.15) of Example 1.5 are linear, whereas Eq. (1.16) of the Example 1.6 is quasi-linear.

1.2 First-order PDEs and Characteristic Strips

1.2.1 *Integral surfaces*

When the variable u depends only on variable x, and if the relation between them is given by an ordinary differential equation (ODE), we need to solve it by integration. The resulting function $u = f(x)$ shows a certain curve in the xu plane, which is called an **integral curve** (or simply an "integral"). This situation is extended to the function u that depends on x and y. When

---

[4]In higher-order PDEs, the term quasi-linear implies that the highest order derivatives of the relevant dependent variables constitute linear combinations.

the function $u = \psi(x, y)$ satisfies the PDE (1.8), it is also called a **solution** or an **integral**. If we consider a three-dimensional space xyz, and plot the value $u = \psi(x, y)$ for the z coordinate, then it gives a certain surface, called an **integral surface**.[5] In the following, we assume that the functions pertaining to the equation are continuously differentiable (*i.e.*, the functions are continuous and their first-order partial derivatives are continuous) in the region under consideration. In this case, the integral surface becomes a smooth surface, and the intuitive geometrical interpretation of Eq. (1.8) is possible, in which the equation $f(x, y, u, p, q) = 0$ provides the information on the gradient p, q of the tangent plane at the point (x, y, u) on that surface.

1.2.2 *Characteristic curves and characteristic strips*

a. Characteristic curves

As a simple example, we consider the first-order linear PDE for the function $u(x, y)$:

$$ap + bq = 0, \quad p = \frac{\partial u}{\partial x}, \quad q = \frac{\partial u}{\partial y}, \tag{1.19}$$

where x, y are the independent variables, and a, b are constants. We assume that a and b are not equal to zero simultaneously ($a^2 + b^2 \neq 0$) whenever we deal with equation of this form, unless otherwise stated. We choose a point P on the integral surface "$u(x, y) = C$ (constant)" where Eq. (1.19) is satisfied. In the neighborhood of P on the integral surface, the relation

$$du = p\,dx + q\,dy = 0$$

holds. By comparing the latter with Eq. (1.19), we obtain[6]

$$\frac{dx}{a} = \frac{dy}{b}, \tag{1.20}$$

---

[5]In general, if a PDE gives the relation among the independent variables $x_1, x_2, \ldots, x_n$ and their function u, as well as the partial derivatives of u with respect to x_i, the function $u = \psi(x_1, x_2, \ldots, x_n)$ that satisfies the equation is called an integral surface in the $(n+1)$-dimensional space with coordinates $x_1, x_2, \ldots, x_n, u$.

[6]A generalization of the above procedure is as follows: If $a_1, a_2, \ldots, a_n$ are functions of independent variables $x_1, x_2, \ldots, x_n$, then the solution of the *first-order linear PDE* for $u(x_1, x_2, \ldots, x_n)$:

$$a_1 \frac{\partial u}{\partial x_1} + a_2 \frac{\partial u}{\partial x_2} + \cdots a_n \frac{\partial u}{\partial x_n} = 0 \tag{*}$$

is shown equivalent to the solution of the *system of ODEs*:

$$\frac{dx_1}{a_1} = \frac{dx_2}{a_2} = \cdots = \frac{dx_n}{a_n}. \tag{**}$$

Equations (**) are called the characteristic differential equations for Eq. (*). If $(n-1)$ independent integrals of the latter $f_1, f_2, \ldots, f_{n-1}$ are obtained, then $u =$

which gives a solution

$$bx - ay = \text{constant}. \tag{1.21}$$

Equation (1.21) describes a line in the xy plane, on which $u(x, y)$ takes the same value. Such a curve[7] is called the **characteristic curve** of Eq. (1.19), and the Eq. (1.20) is called the **characteristic differential equation**.

If we consider the above-mentioned relation in the xyz space with u in the z direction, the equation "$u(x, y) = C$ (constant)" gives an integral surface in this space, whereas Eq. (1.21) gives a plane extending to infinity parallel to the z axis. The intersection of these two surfaces gives an integral curve. Since u takes the same value on the characteristic curve, both the characteristic curves on the integral surface and their projections to the xy plane are given by Eq. (1.21).[8] Geometrically, Eq. (1.19) provides a relation that the scalar product $\boldsymbol{t}_0 \cdot \boldsymbol{n}_0 = 0$, which implies that $\boldsymbol{t}_0$ and $\boldsymbol{n}_0$ are perpendicular to each other. Here, $\boldsymbol{t}_0 \propto (a, b)$ is the vector tangent to the characteristic curve Eq. (1.21) that is projected to the xy plane, whereas vector $\boldsymbol{n}_0 \propto (p, q)$ is the normal vector $\boldsymbol{n} \propto \nabla u$ at point P projected to the xy plane. (Note that we adopt the thick italics like $\boldsymbol{v}$ in this textbook to show the vector quantity of magnitude v with specified direction.)

For confirmation, we consider what will occur if we take new variables $\tilde{\eta}$ in the $\boldsymbol{t}_0$ direction, $\tilde{\xi}$ in the $\boldsymbol{n}_0$ direction, and change the independent variables x, y to $\tilde{\xi}, \tilde{\eta}$, such that

$$\tilde{\xi} = bx - ay, \quad \tilde{\eta} = ax + by.$$

In this transformation, the relations

$$p = \frac{\partial u}{\partial x} = \frac{\partial u}{\partial \tilde{\xi}} \frac{\partial \tilde{\xi}}{\partial x} + \frac{\partial u}{\partial \tilde{\eta}} \frac{\partial \tilde{\eta}}{\partial x} = b \frac{\partial u}{\partial \tilde{\xi}} + a \frac{\partial u}{\partial \tilde{\eta}}, \quad q = \frac{\partial u}{\partial y} = -a \frac{\partial u}{\partial \tilde{\xi}} + b \frac{\partial u}{\partial \tilde{\eta}},$$

hold, so that Eq. (1.19) becomes

$$\frac{\partial u(\tilde{\xi}, \tilde{\eta})}{\partial \tilde{\eta}} = 0, \quad \text{or} \quad u = u(\tilde{\xi}) = u(bx - ay).$$

$\psi(f_1, f_2, \ldots, f_{n-1})$ gives the general solution of Eq. (*), where ψ is an arbitrary function of $(n-1)$ variables.

[7]Although it is a line in the present example, it is given by a curve in general.

[8]For this reason, the term "characteristic curve" may refer to the curve on the integral surface, or may refer to the curve that is projected to the independent variable space. Although the latter is also called the "characteristic basic curve", we do not rigorously distinguish between the two, and refer it to the "characteristic curve in the xy plane" if necessary. Similar situations are found in cartography. A contour line joins points of equal elevation, which is a 2D projection of the real (3D) topographic map or integral surface. The former corresponds to the characteristic basic curve, whereas the latter corresponds to the characteristic curve or integral curve.

This relation shows that u depends only on the variable $\tilde{\xi}$, and does not depend on the other variable $\tilde{\eta}$. In other words, the first-order PDE for the function of two independent variables is reduced to the ODE by the use of a new variable $\tilde{\xi}$. The value of u at a point $\tilde{\xi} = \tilde{\xi}_0$ (initial value $u(\tilde{\xi}_0)$) is the same at any point on the curve that satisfies the equation $bx - ay = \tilde{\xi}_0$, from which we can interpret that the initial value propagates along this curve in the xy plane.[9] Note that the initial value $\tilde{\xi}_0$ is specified by a variable that is independent of $\tilde{\xi}$. The latter variable may be any variable along the curve that intersects the curve "$\tilde{\xi} =$ constant" (see Fig. 1.1).

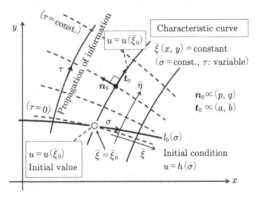

Fig. 1.1   Schematic picture of the characteristic curves (σ and τ are the coordinates along the curve l_0 and the curve "$\tilde{\xi} =$ constant", respectively).

We shall here summarize the features of the **characteristic curve**:

• It gives the path along which a given quantity (or information) propagates subject to the given PDE, so that the value of u is the same along this curve.

• The vector tangent to this curve is parallel to the direction in which the information propagates, and is determined by the characteristic differential equation. The latter is a first-order ODE, which is always solvable.

• If we give the initial value continuously on the curve that intersects a characteristic curve, we obtain a family of characteristic curves in the xyu space, which forms an integral surface. Characteristic curves are perpendicular to the normal vector on the integral surface, the relation of which is given by the first-order linear PDE. (Note that the "characteristic curves in the xy plane" are those that are projected on the xy plane.)

---

[9]Given one of the variables, *e.g.*, $x = x_1, x_2$, then the other variables y_1, y_2 on this curve are determined by the relation (1.21), and the values of u at these points are the same, which implies that the value ("information") of u at (x_1, y_1) propagates to the point (x_2, y_2) along this curve.

Example 1.7 (Plane wave). We shall follow the statements thus far developed using a simple example:

$$u(x,t) = A\cos(kx - \omega t + \delta). \tag{1.22}$$

The function (1.22) fulfills the PDE

$$\omega\frac{\partial u}{\partial x} + k\frac{\partial u}{\partial t} = \omega p + kq = 0, \tag{1.23}$$

where the variable y in the Eq. (1.19) is replaced by t, and (a,b) corresponds to (ω, k). Accordingly, the characteristic curve (1.21) is given by

$$kx - \omega t = \text{constant.} \tag{1.24}$$

The direction of the vector $\boldsymbol{t}_0$ tangent to this curve is $\boldsymbol{t}_0 \propto (a,b) \propto (\omega, k)$, whereas $\boldsymbol{n}_0 \propto (p,q) \propto (-kA\sin\phi, \omega A\sin\phi)$, where we put $\phi = kx - \omega t + \delta$ for brevity. Equation (1.23) shows that $\boldsymbol{t}_0$ is perpendicular to (p,q). To see the physical meaning, we take the x axis along a line in space, the z axis for u, and the third axis for t (time). Then Eq. (1.22) describes the wave that propagates in the x direction with a speed $v = \omega/k$, where A and ϕ are the wave amplitude and phase, respectively. Note that, if we move with a particular characteristic curve (1.24), the wave height does not change, so that the wave looks stationary. The integral surface in the xtz space is covered by a family of characteristic curves that are parallel to Eq. (1.24), and hence is given by a cylindrical surface with its generating line in this direction (see Fig. 1.2).

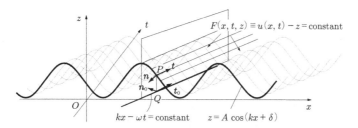

Fig. 1.2 Integral surface and characteristic curves of the plane wave.

The light wave propagating in a three-dimensional space can be interpreted, on the one hand, as the propagation of the wave front of a two-dimensional displacement with the same phase. On the other hand, the same physical phenomenon is interpreted as a one-dimensional path, described by the characteristic curve, on which the displacement (or some type of information) at a marked point propagates. The latter is the ray of light and plays a basic role in geometrical optics (or ray optics).

Note again that the characteristic curves in Example 1.7 happen to be *lines*. The direction of propagation of the wave, however, varies in general when the wave propagates in an inhomogeneous medium, and/or the amplitude of displacement varies spatially. In these cases, characteristic curves are given by *curves*. We show such an example in the following.

Example 1.8. Obtain the solution of the first-order PDE for $u(x, y)$ that depends on two independent variables x and y:

$$p + yq = 0. \tag{1.25}$$

As stated before, the characteristic differential equation is

$$\frac{\mathrm{d}x}{1} = \frac{\mathrm{d}y}{y},$$

which yields the solution

$$y = C\,\mathrm{e}^x, \qquad (C: \text{ arbitrary constant}).$$

The latter is a characteristic *curve*. Owing to the presence of an arbitrary constant C, it gives a "family of curves", and the general solution is given by

$$u(x, y) = \psi(y\,\mathrm{e}^{-x}),$$

where ψ is an arbitrary function. $\triangleleft$

We shall now consider the *first-order quasi-linear PDE* in general. We assume, as before, that the unknown function u depends on two independent variables x and y, and that the integral surface is given by "$F(x, y, z) \equiv u(x, y) - z =$ constant." As shown in Fig. 1.3, the direction of the vector $\boldsymbol{n}$ normal to the integral surface at a point $P(x, y, z)$ is given by[10]

$$\boldsymbol{n} \propto \left(\frac{\partial F}{\partial x}, \frac{\partial F}{\partial y}, \frac{\partial F}{\partial z} \right) = \left(\frac{\partial u}{\partial x}, \frac{\partial u}{\partial y}, -1 \right) = (p, q, -1). \tag{1.26}$$

---

[10]Consider a point P' on the surface $F =$ constant, which is infinitesimally displaced in an arbitrary direction $\mathrm{d}\boldsymbol{r} = (\mathrm{d}x, \mathrm{d}y, \mathrm{d}z)$ from the point P. The value of F at P' is the same as the one at P, so that (as far as the approximation up to the first order of $|\mathrm{d}\boldsymbol{r}|$ is concerned)

$$0 = F(x + \mathrm{d}x, y + \mathrm{d}y, z + \mathrm{d}z) - F(x, y, z) = \frac{\partial F}{\partial x}\mathrm{d}x + \frac{\partial F}{\partial y}\mathrm{d}y + \frac{\partial F}{\partial z}\mathrm{d}z = \nabla F \cdot \mathrm{d}\boldsymbol{r},$$

and hence ∇F and $\mathrm{d}\boldsymbol{r}$ are perpendicular to each other. Since $\mathrm{d}\boldsymbol{r}$ is an arbitrary displacement vector in the neighborhood P and is on the surface $F =$ constant, the above relation implies that ∇F is perpendicular to the tangent plane at point P, so that ∇F has the same direction as the normal vector $\boldsymbol{n}$.

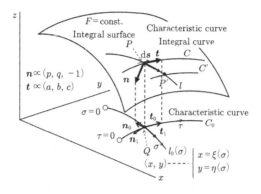

Fig. 1.3   Integral surface and characteristic curve.

If we consider the vector $t \propto (a, b, c)$ at a point P that is perpendicular to n, then we have

$$n \cdot t = 0 \quad \text{i.e.,} \quad ap + bq - c = 0,$$

which implies that

t is tangent to the integral surface at a point P.

We then generate a curve C on the surface that connects the neighboring points successively in the t direction, such that it fulfills the equation

$$a(x, y, u)\, p + b(x, y, u)\, q = c(x, y, u). \tag{1.27}$$

Since Eq. (1.27) is of the same form as the Eq. (1.17) mentioned in the definition of the quasi-linear PDEs, the curve C becomes "an integral curve" of the PDE (1.17), if we choose

$$\frac{P(x, y, u)}{a(x, y, u)} = \frac{Q(x, y, u)}{b(x, y, u)} = \frac{R(x, y, u)}{c(x, y, u)}.$$

On the curve C, the line element $ds = (dx, dy, du)$ is parallel to t:

$$\frac{dx}{a(x, y, u)} = \frac{dy}{b(x, y, u)} = \frac{du}{c(x, y, u)}, \tag{1.28}$$

and hence the integral curve of the PDE (1.17) is determined by solving the equations

$$\frac{dx}{P(x, y, u)} = \frac{dy}{Q(x, y, u)} = \frac{du}{R(x, y, u)}. \tag{1.29}$$

As mentioned previously on solving Eq. (1.20), Eq. (1.29) is the **characteristic differential equation**, and the vector (a, b, c) or (P, Q, R) is the

characteristic direction. Since Eqs. (1.29) are a system of two differential equations, two surfaces $F_1(x, y, z) = C_1$ and $F_2(x, y, z) = C_2$ (C_1, C_2 : constants) are obtained as their solutions, and their intersection C gives the integral curve. The latter provides the **characteristic curve.**

On the smooth surface, a point P' exists in the neighborhood of point P, at which the relation $\boldsymbol{n} \cdot \boldsymbol{t} = 0$, *i.e.*, Eq. (1.27) holds. Then, starting from the point P', a new solution curve C' is created similarly by successively connecting the direction $\boldsymbol{t}$ (*i.e.*, the vector with direction (a, b, c) or (P, Q, R)). If we choose such a starting point on a certain curve l that intersects the curve C at point P (see Fig. 1.3), and repeat the abovementioned procedures over all starting points on the curve l, we obtain a family of integral curves, which becomes the integral surface being sought for the given PDE.

In summary, the first-order quasi-linear PDE (1.27) gives the relation required for the integral surface in which the normal vector $\boldsymbol{n} \propto (p, q, -1)$ at any point on the surface is perpendicular to the tangential vector $(a, b, c) \propto \boldsymbol{t}$ at that point. The curve that connects the neighboring points successively along this direction forms a characteristic curve, which holds true for its immediate neighboring points, so that the integral surface is a union of the characteristic curves. Surprisingly, the problem of solving the first-order general PDE is reduced to solving a system of ODEs.

Remark 1.1. In the above explanation, we interpreted the characteristic curve as an intersection of the two surfaces that satisfies the system of two differential equations (1.28) [or (1.29)]. We can add another view. If we rewrite Eq. (1.28) as

$$\frac{\mathrm{d}x}{a(x, y, u)} = \frac{\mathrm{d}y}{b(x, y, u)} = \frac{\mathrm{d}u}{c(x, y, u)} = \mathrm{d}\tau,$$

by introducing a new parameter τ (note that the latter is introduced as an another independent variable), then we have a system of three equations:

$$\frac{\mathrm{d}x}{\mathrm{d}\tau} = a(x, y, u), \quad \frac{\mathrm{d}y}{\mathrm{d}\tau} = b(x, y, u), \quad \frac{\mathrm{d}u}{\mathrm{d}\tau} = c(x, y, u). \tag{1.30}$$

Equations (1.30) are first-order ODEs, which are integrated to have the solution curves:

$$x = x(\tau), \quad y = y(\tau), \quad u = u\big(x(\tau), y(\tau)\big). \tag{1.31}$$

On this curve, the relation

$$\frac{\mathrm{d}u}{\mathrm{d}\tau} = \frac{\partial u}{\partial x}\frac{\mathrm{d}x}{\mathrm{d}\tau} + \frac{\partial u}{\partial y}\frac{\mathrm{d}y}{\mathrm{d}\tau}, \qquad i.e., \qquad c = pa + qb$$

holds. The latter is the given PDE (1.27) itself, so that Eq. (1.31) gives the characteristic curve, where the parameter τ is regarded as a coordinate along the line (see Fig. 1.3). Equation (1.30) is also called the characteristic differential equation.

As an example in physics, we may look at the velocity field of a fluid. The latter is the vector field v that shows the velocity components in the fluid, which is determined by the relevant PDE (*e.g.*, Navier–Stokes equation). If we denote the "direction vector field" (a, b, c), and consider the marker particle that moves with the flow, then its path is governed by Eq. (1.30). If we regard τ as the time, and x as the position in the x direction, the quantity a corresponds to the velocity component in the x direction. The path (or orbit) that follows the marker particle with time is found to be the characteristic curve starting from the initial position. ◁

b. Characteristic strips

We shall now consider a more general *first-order PDE* with two independent variables

$$f(x, y, u, p, q) = 0, \qquad p = \frac{\partial u}{\partial x}, \qquad q = \frac{\partial u}{\partial y}. \tag{1.8}$$

This equation provides a relation among the five data (x, y, u, p, q). From a geometrical perspective, it is regarded as a requirement to the directions (p, q) at the point (x, y, u) on the integral surface. In the linear or quasi-linear PDEs, we can easily obtain the direction vector at every point (x, y, u). How is the case in general (including non-linear) first-order PDEs?

To see this, we first consider a point $P(x_0, y_0, u_0)$ on a certain surface element dS. Then the unit normal vector n at P is given by

$$n = \frac{(p, q, -1)}{\sqrt{p^2 + q^2 + 1}}, \tag{1.32}$$

and the equation for the tangent plane passing through the point P is

$$u - u_0 = p(x - x_0) + q(y - y_0), \tag{1.33}$$

where x, y and u are the running coordinates. As we have seen in Example 1.5, Eq. (1.33) generally describes a cone with a vertex at P (sometimes referred to as a "normal cone"). At the same time, the "**envelope**"[11] of the tangent planes passing through the point P also forms a cone, which

---

[11]Suppose $f(x, y, C) = 0 \cdots (*)$, where C is an arbitrary constant (or parameter). This relation gives a one-parameter family of solutions in a two-dimensional space. With a

may be called a "tangent cone". Since the normal vector takes any one of an infinite number of directions, the associated surface element dS is not uniquely determined in general PDEs. Then how is the particular generator of the tangent cone that belongs to the integral surface determined?

Hereafter, we assume that the point P belongs to the integral surface *i.e.*, it satisfies Eq. (1.8), or $f(x_0, y_0, u_0, p, q) = 0$. The latter gives a functional relation between p and q, so that Eq. (1.33) shows a one-parameter family of lines passing through the point P, *i.e.*, a conical surface that consists of the normal vectors. The associated conical surface (tangent cone) is called the **Monge cone**, and any tangent plane in contact with the Monge cone becomes a possible integral surface. If we choose p for the control parameter, so that q may be given by $q = g(p)$, Monge cone is given by eliminating p (and accordingly q) using Eq. (1.33) and the equation $\partial u/\partial p = 0$ as shown in footnote 11. Namely, the set of equations

$$du = p \, dx + g(p) \, dy, \quad dx + g'(p) dy = 0,$$

gives the Monge cone, where we denote $dx = x - x_0$, $dy = y - y_0$, and $du = u - u_0$, for brevity. Instead of the latter formulation, however, we calculate the above procedure by keeping the symmetry of p and q. Then the relation between p and q is given by $df = 0$, *i.e.*,

$$\frac{\partial f}{\partial p} dp + \frac{\partial f}{\partial q} dq = 0. \tag{1.34}$$

One of the requirement for the Monge cone is the same as before, except for the abbreviated notation, so that Eq. (1.33) follows

$$du = p \, dx + q \, dy, \tag{1.33}'$$

whereas the other requirement for the Monge cone becomes

$$dp \, dx + dq \, dy = 0. \tag{1.35}$$

---

change of C by dC, $(|dC| \ll |C|)$, the relation (*) also holds, so that we have

$$f(x, y, C + dC) = f(x, y, C) + \frac{\partial f}{\partial C} dC + \cdots = 0 \quad \rightarrow \quad \frac{\partial f}{\partial C} = 0. \cdots (**)$$

Equations (*) and (**) give the point(s) in common. Similar solution points are generated by changing C, whose union constitutes the "envelop". The mathematical procedure for doing this is to eliminate C using Eqs. (*) and (**). Furthermore, if the relation $f(x, y, u, C) = 0$ is given, the envelop is also determined by eliminating C with the use of another equation $\partial f/\partial C = 0$. In this case, the solution gives a surface in the xyu space for each C, whose intersection with different values of C gives a curve. Then the union of these curves forms an envelop surface. (In the above, the smoothness of f and the real parameter C are assumed.)

From Eqs. (1.34), (1.35), and (1.33)$'$, we obtain the equations

$$\frac{\mathrm{d}x}{\dfrac{\partial f}{\partial p}} = \frac{\mathrm{d}y}{\dfrac{\partial f}{\partial q}} = \frac{\mathrm{d}u}{p\left(\dfrac{\partial f}{\partial p}\right) + q\left(\dfrac{\partial f}{\partial q}\right)}. \tag{1.36}$$

Consequently, we reached the conclusion that, with a choice of normal vector at P, and hence the Monge cone being determined, a particular vector on the tangent plane that touches the latter specifies the characteristic direction $(\mathrm{d}x, \mathrm{d}y, \mathrm{d}u)$, which is given by Eq. (1.36) (see Fig. 1.4(a)).

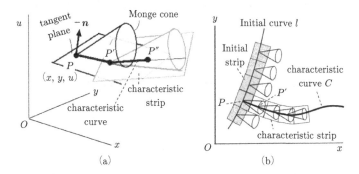

Fig. 1.4   Monge cones, characteristic curves, and characteristic strips.

Second, we move from point P to P' with an infinitesimally small distance apart along the line of contact that is determined by the previous procedure (see Fig. 1.4(a)). The same process is repeated to generate a successively connected curve, which leads to the **characteristic curve**. Here, consecutive infinitesimal planes in contact with the Monge cone at the new positions are arranged one-dimensionally along the characteristic curve. The latter constitutes a strip of an integral surface, which is called the **characteristic strip**.

Finally, if we perform the above-mentioned procedures by choosing the initial positions along the line l that intersects the characteristic curve at P, then we have similar characteristic strips, which are put together to generate the whole integral surface that satisfies the initial condition (see Fig. 1.4(b)).

For further details, see *e.g.*, [John (1982)]. Note that Eq. (1.36) is a prototype of the Lagrange–Charpit method, which is to be fully described in §§1.4.4.

Example 1.9. Show how the *quasi-linear* PDE is expressed in the form of Eq. (1.36).

A quasi-linear PDE (1.27) is given in the form $f \equiv ap + bq - c = 0$, from which we have

$$\frac{\partial f}{\partial p} = a, \quad \frac{\partial f}{\partial q} = b, \quad \left(\frac{\partial f}{\partial p}\right)p + \left(\frac{\partial f}{\partial p}\right)q = ap + bq = c.$$

Substituting the above expressions for Eq. (1.36), we have

$$\frac{dx}{a} = \frac{dy}{b} = \frac{du}{c},$$

which is the same as Eq. (1.28). A quasi-linear PDE (1.27) is a special case, in which the Monge cone at any point P degenerates into the line with direction (a, b, c) through P. ◁

1.2.3 *Solution of the first-order quasi-linear PDEs*

In this subsection, we consider the general solution of the first-order quasi-linear PDEs, or the Lagrange's PDEs:

$$P(x, y, u)\, p + Q(x, y, u)\, q = R(x, y, u). \tag{1.17}$$

To do so, we first seek the solution of the characteristic equations, or the auxiliary equations

$$\frac{dx}{P(x, y, u)} = \frac{dy}{Q(x, y, u)} = \frac{du}{R(x, y, u)}. \tag{1.29}$$

If the latter solutions:

$$f(x, y, u) = C_1, \quad g(x, y, u) = C_2, \quad (C_1, C_2 : \text{arbitrary constants}), \tag{1.37}$$

are obtained, then

$$\varphi(f, g) = 0, \quad \text{or} \quad g = \psi(f) \tag{1.38}$$

gives the general solution, where φ and ψ are arbitrary functions. ◁

We shall show that $\varphi(f, g) = 0$ is indeed the general solution of the Eq. (1.17). Since each of the relations in Eq. (1.37) gives a smooth surface $f = C_1$ and $g = C_2$, we have $df = 0$ and $dg = 0$ on the respective surfaces, *i.e.*,

$$f_x dx + f_y dy + f_u du = 0, \quad g_x dx + g_y dy + g_u du = 0.$$

Together with Eq. (1.29), we have

$$P f_x + Q f_y + R f_u = 0, \quad P g_x + Q g_y + R g_u = 0,$$

from which we have, after some calculation,

$$\frac{P}{\partial(f,g)/\partial(y,u)} = \frac{Q}{\partial(f,g)/\partial(u,x)} = \frac{R}{\partial(f,g)/\partial(x,y)}. \tag{1.39}$$

We have already shown in Example 1.6 that a surface given by $\varphi(f,g) = 0$ satisfies the following equation:

$$p\,\frac{\partial(f,g)}{\partial(y,u)} + q\,\frac{\partial(f,g)}{\partial(u,x)} = \frac{\partial(f,g)}{\partial(x,y)}. \tag{1.40}$$

By inserting Eq. (1.39) into Eq. (1.40), Eq. (1.17) is reproduced. Since our solution satisfies the given first-order PDE and includes one arbitrary function, it gives the general solution. ◁

Example 1.10. In our previous example (Example 1.7), the auxiliary equations (1.29) are

$$\frac{\mathrm{d}x}{\omega} = \frac{\mathrm{d}t}{k} = \frac{\mathrm{d}u}{0},$$

from which we have[12]

$$f(x,t,u) \equiv kx - \omega t = C_1, \qquad g(x,t,u) \equiv u = C_2,$$

where C_1 and C_2 are arbitrary constants. By eliminating the constants C_1 and C_2, we have the general solution

$$\varphi(kx - \omega t, u) = 0, \qquad \text{or} \qquad u = \psi(kx - \omega t),$$

where φ and ψ are arbitrary functions. The precise forms of these functions, (including the amplitude A and phase ϕ) are determined by the initial and/or boundary conditions. ◁

Example 1.11. Obtain the general solution of the PDE

$$xp + yq = u.$$

This equation has the same form as Eq. (1.17), where P, Q, and R correspond to x, y, and u, respectively. The auxiliary equations (1.29) are

$$\frac{\mathrm{d}x}{x} = \frac{\mathrm{d}y}{y} = \frac{\mathrm{d}u}{u},$$

---

[12]This expression does not mean that $\mathrm{d}u$ is divided by "0". If, for example, we equate the first and second equations to the last equation, it gives $\mathrm{d}u/\mathrm{d}x = \mathrm{d}u/\mathrm{d}t = 0$, respectively, from which we have $u = $ constant.

from which we have

$$\frac{x}{y} = C_1, \qquad \frac{u}{y} = C_2, \qquad (C_1, C_2 : \text{arbitrary constants}),$$

and hence, the general solution is given by Eq. (1.38):

$$\varphi\left(\frac{x}{y}, \frac{u}{y}\right) = 0, \quad \text{or} \quad u = y\,\psi\left(\frac{x}{y}\right), \quad \left[\text{or} \quad u = x\,\tilde{\psi}\left(\frac{y}{x}\right)\right],$$

where φ and ψ [or, $\tilde{\psi}$] are arbitrary functions. Note that the first equation can also be rewritten, using another arbitrary function $\tilde{\varphi}$, as

$$\tilde{\varphi}\left(\frac{u}{x}, \frac{u}{y}\right) = 0,$$

taking into account that $x/y = (u/y)/(u/x)$. The latter agrees with the equation describing the conical surface (1.14).   ◁

The method of solution developed in this subsection can be extended to the partial differential equations depending on many more variables. Consider the following PDE:

$$P_1\frac{\partial u}{\partial x_1} + P_2\frac{\partial u}{\partial x_2} + \cdots + P_n\frac{\partial u}{\partial x_n} = R, \qquad (1.41)$$

where u is a function of $x_1, x_2, \ldots, x_n$, whereas P_i $(i = 1, \ldots, n)$ and R are (continuously differentiable) functions of $x_1, x_2, \ldots, x_n, u$, and not simultaneously zero. The corresponding auxiliary equations (characteristic differential equations) are

$$\frac{dx_1}{P_1} = \frac{dx_2}{P_2} = \cdots = \frac{dx_n}{P_n} = \frac{du}{R}. \qquad (1.42)$$

Using n independent solutions of the latter

$$f_i(x_1, x_2, \ldots, x_n, u) = C_i, \qquad (i = 1, 2, \ldots, n), \qquad (1.43)$$

we obtain the general solution

$$\varphi(f_1, f_2, \ldots, f_n) = 0, \qquad (1.44)$$

where C_i are arbitrary constants, and φ is an arbitrary function.

1.3   Initial-value Problems of the First-order PDEs

a. Initial-value problem (Cauchy's problem)

In the ODEs, we call the initial-value problems as those requiring a solution in which the dependent variable fulfills the given conditions involving the values and/or the derivatives of the dependent variables at an initial instant. Similarly, the problems to obtain the solution of the PDEs that fulfills the given "function" at an initial instant are generally called the **initial-value problems**, or the **Cauchy's problems**. From a geometrical viewpoint, it is the problem of obtaining an integral surface that passes a given curve or matches the gradient there at an initial instant. Note that the term "initial" need not refer exactly to the time at $t = 0$. If the condition is given at some instant $t = t_0$, then the latter time is regarded as the "initial" time. If relevant variables are spatial, and the conditions are given *e.g.*, at a certain position $x = 0$, these conditions serve as the initial values. The latter problems are instead called the "boundary-value problems", in which the initial-value may be regarded as the boundary-value. Thus, in the typical example of the first-order PDEs, with the dependent variable u of the function of x and y:

$$f(x, y, u, p, q) = 0, \qquad p = \frac{\partial u}{\partial x}, \qquad q = \frac{\partial u}{\partial y}, \tag{1.8}$$

the Cauchy's problem will be to obtain the solution of $u(x, y)$ for $x \geq 0$, that satisfies the initial condition at $x = 0$:

$$u(0, y) = h(y). \tag{1.45}$$

Of course, the continuity and differentiability of $h(y)$ are assumed.

b. Initial-value problems and series expansions

One of the methods for solving the initial-value problems is to consider the power series expansion around a point $x = 0$:

$$u(x, y) = u(0, y) + \left(\frac{\partial u}{\partial x}\right)_0 x + \frac{1}{2} \left(\frac{\partial^2 u}{\partial x^2}\right)_0 x^2 + \cdots . \tag{1.46}$$

The subscripts "0" at the parentheses on the right-hand side of Eq. (1.46) are intended to evaluate the pertaining quantities at $x = 0$. If we can solve p from the given equation Eq. (1.8), and is expressed by

$$p \equiv \frac{\partial u}{\partial x} = g(x, y, u, q), \tag{1.47}$$

then g is a known function, and we can calculate

$$\left(\frac{\partial u}{\partial x}\right)_0 = g\big(0, y, h(y), h'(y)\big),$$

$$\left(\frac{\partial^2 u}{\partial x \partial y}\right)_0 = \frac{d}{dy} g\big(0, y, h(y), h'(y)\big) = \left(\frac{\partial q}{\partial x}\right)_0,$$

where the differentiability of g is assumed. By performing partial differentiation of Eq. (1.47) with respect to x, we can calculate the initial value

$$\left(\frac{\partial^2 u}{\partial x^2}\right)_0 = \left(\frac{\partial g}{\partial x}\right)_0 + \left(\frac{\partial g}{\partial u}\right)_0 \left(\frac{\partial u}{\partial x}\right)_0 + \left(\frac{\partial g}{\partial q}\right)_0 \left(\frac{\partial q}{\partial x}\right)_0.$$

Similarly, we can calculate the initial values of $\partial^i u / \partial x^i$, $(i = 1, 2, 3, \cdots)$ by successive partial differentiations. In this way, we can obtain the solution that fulfills the initial condition within the radius of convergence of the series.

c. Initial-value problems and characteristic curves

In this subsection, we show the method of characteristic curves to solve the initial-value problems. We consider the general case in which the initial conditions are given on the curve l_0 that depends on x and y, in addition to the case that is not as simple as those given at $x = 0$. The curve l_0 may be specified as $y = g(x)$, but here we adopt the expression $x = x(\sigma)$ and $y = y(\sigma)$ in terms of the parameter σ. In this case, u is also expressed as a function of σ: $u = u\big(x(\sigma), y(\sigma)\big)$, and the initial condition $u = h(\sigma)$ is given on this curve (see Figs. 1.1 and 1.3). At the point $P(x, y, u)$ on the initial curve l (where l is the curve in the xyu space corresponding to l_0), the normal vector ∇u and the tangent plane are specified on the integral surface, and PDE (1.8) gives the relation between the gradient p and q in the x and y directions, respectively. As shown in §§1.2.2, these data specify the Monge cone, and characteristic strips and characteristic curves are determined in the direction of the neighboring infinitesimal tangent planes (see Fig. 1.4). By shifting the point P along the curve l, we can obtain the solution of Eq. (1.8) starting from the newly chosen initial position, which yields the entire solution of the initial-value problem.

Remark 1.2. In order to obtain the unique solution for the initial-value problems, the curve l_0 that specifies the initial condition and the characteristic curve C_0 must intersect only once in the region under consideration. If we assume that the initial curve l_0 and the characteristic curve C_0 are specified in terms of the parameters σ and τ, respectively, so that x and y are expressed by $x = x(\sigma, \tau)$ and $y = y(\sigma, \tau)$, then the previously-mentioned condition is described as $J \neq 0$, where J is the functional determinant (or Jacobian):

$$J \equiv \frac{\partial x}{\partial \sigma}\frac{\partial y}{\partial \tau} - \frac{\partial x}{\partial \tau}\frac{\partial y}{\partial \sigma} = \frac{\partial(x,y)}{\partial(\sigma,\tau)}. \tag{1.48}$$

◁

For the detailed proof on the existence and uniqueness of the solution, consult *e.g.*, [John (1982)]. The initial-value problems in the second-order PDEs (Cauchy's problem) require another condition besides the condition (1.45), which will be shown in §§2.2.2 of Chapter 2.

1.4 Complete Integrals

1.4.1 *Complete solutions, general solutions, and singular solutions*

a. Complete solutions

In the first-order PDEs for u that depend on two variables x and y:

$$f(x, y, u, p, q) = 0, \qquad p = \frac{\partial u}{\partial x}, \qquad q = \frac{\partial u}{\partial y}, \tag{1.8}$$

a solution, such as

$$\varphi(x, y, u, C_1, C_2) = 0, \tag{1.49}$$

or

$$u = \psi(x, y, C_1, C_2), \tag{1.50}$$

in which as many arbitrary constants as the number of independent variables (here two constants C_1 and C_2) are included, is called a **complete solution** or **complete integral**.

b. General solutions, particular solutions, and singular solutions

Once we obtain the complete solution, we can derive other types of solutions. We show this using the previous example. We first differentiate Eq. (1.49) with respect to x and y assuming that C_1 and C_2 are constants, which yields

$$\frac{\partial \varphi}{\partial x} + p\frac{\partial \varphi}{\partial u} = 0, \qquad \frac{\partial \varphi}{\partial y} + q\frac{\partial \varphi}{\partial u} = 0. \tag{1.51}$$

If we determine C_1 and C_2 from these equations, and substitute them for Eq. (1.49), then we obtain the original equation (1.8). Next, we differentiate

Eq. (1.49) with respect to x and y assuming that C_1 and C_2 are functions of x and y, which gives[13]

$$\frac{\partial \varphi}{\partial C_1}\frac{\partial C_1}{\partial x} + \frac{\partial \varphi}{\partial C_2}\frac{\partial C_2}{\partial x} = 0, \qquad \frac{\partial \varphi}{\partial C_1}\frac{\partial C_1}{\partial y} + \frac{\partial \varphi}{\partial C_2}\frac{\partial C_2}{\partial y} = 0. \qquad (1.52)$$

We consider the following three cases that fulfill the Eqs. (1.49) and (1.52):

- (Case 1) C_1 and C_2 are constants: This gives a complete solution.
- (Case 2) Both $\partial \varphi / \partial C_1 = 0$ and $\partial \varphi / \partial C_2 = 0$ hold: In this case, it is trivial that Eq. (1.52) holds. Since φ does not depend on C_1 nor on C_2, we can solve u, C_1, C_2 as functions of x and y, so that arbitrary constants are not included. The latter type of solution is called a **singular solution** or **singular integral**.
- (Case 3) Either $\partial \varphi / \partial C_1 = 0$ or $\partial \varphi / \partial C_2 = 0$ holds: Equations (1.52) are a system of linear equations for $\partial \varphi / \partial C_1$ and $\partial \varphi / \partial C_2$, so that the present conditions are allowed only if $J = 0$, where J is the Jacobian $J \equiv \partial(C_1, C_2)/\partial(x, y)$. In this case, C_1 and C_2 are not independent to each other, and have a certain functional relation (see the footnote of Example 1.3). For example, if $C_2(x, y) = g\big(C_1(x, y)\big)$ is given, we can calculate

$$\frac{\partial \varphi}{\partial C_1} + \frac{\partial \varphi}{\partial C_2}g'(C_1) = 0,$$

so that, together with the relation $\varphi = 0$, we can eliminate C_1 and C_2. The resulting solution, however, allows one arbitrary function g. This type of solution, *i.e.*, the solution that includes as many arbitrary functions as the number of the order of PDE (in the present first-order PDE, one arbitrary function g being included) is called a **general solution** or **general integral**. Furthermore, the solution obtained by assigning particular conditions in the general solution is called a **particular solution** or **particular integral**.[14]

---

[13]In general, if we assume that u, C_1, C_2 are functions of x, y, and calculate the partial derivatives of Eq. (1.49) with respect to x and y, then we have

$$\frac{\partial \varphi}{\partial x} + \frac{\partial \varphi}{\partial u}p + \frac{\partial \varphi}{\partial C_1}\frac{\partial C_1}{\partial x} + \frac{\partial \varphi}{\partial C_2}\frac{\partial C_2}{\partial x} = 0, \qquad \frac{\partial \varphi}{\partial y} + \frac{\partial \varphi}{\partial u}q + \frac{\partial \varphi}{\partial C_1}\frac{\partial C_1}{\partial y} + \frac{\partial \varphi}{\partial C_2}\frac{\partial C_2}{\partial y} = 0.$$

For a further calculation, we do not lose generality if we impose a condition on the arbitrary constants C_1, C_2 such that they fulfill the relation (1.51), from which we obtain Eq. (1.52). This procedure is similar to the method of variation of constants (variation of parameters) in the second-order ODEs.

[14]The solution in which some numerical values are assigned to the constants C_1 and C_2 in the complete solution is also a particular solution, because it does not includes arbitrary constants.

Example 1.12. The PDEs dealt with in Eq. (1.7) of Example 1.1 [or in Eq. (1.11) of Example 1.3]:

$$yp - xq = 0 \tag{1.7}$$

is of a Lagrange-type PDE. The solution of the latter is obtained by the method shown in §§1.2.3., *i.e.*, by solving the auxiliary equations

$$\frac{\mathrm{d}x}{y} = -\frac{\mathrm{d}y}{x} = \frac{\mathrm{d}u}{0},$$

we have $f \equiv x^2 + y^2 = C_1$, $g \equiv u = C_2$, where C_1 and C_2 are arbitrary constants. By eliminating C_1 and C_2 from the above relations, we have $\varphi(u, x^2 + y^2) = 0$ or $u = \psi(x^2 + y^2)$, where φ and ψ are arbitrary functions. These solutions are *general solutions* of the PDE (1.7).

Note that if we impose the condition, *e.g.*,

$$u = \pm c\sqrt{1 - \frac{x^2}{a^2}} \quad \text{at} \quad y = 0,$$

then we can determine the arbitrary function ψ as follows:

$$u = \psi(x^2) = \pm c\sqrt{1 - \frac{x^2}{a^2}}, \quad i.e., \quad \psi(\xi) = \pm c\sqrt{1 - \frac{\xi}{a^2}}.$$

With this function, we obtain the solution as a function of x and y:

$$u = \psi(x^2 + y^2) = \pm c\sqrt{1 - \frac{x^2 + y^2}{a^2}}, \quad i.e., \quad \frac{x^2 + y^2}{a^2} + \frac{u^2}{c^2} = 1,$$

which gives the ellipsoid of revolution shown in Eq. (1.4) of Example 1.1. The latter is a *particular solution* of the PDE (1.7) that fulfills the given condition. ◁

Example 1.13. Obtain the solution of the PDE

$$u = xp + yq + p^2 + q^2. \tag{1.53}$$

This is a type of PDE called the **Clairaut's equation**. This equation is also written in the form of Eq. (1.8) as

$$f(x, y, u, p, q) \equiv u - xp - yq - p^2 - q^2 = 0.$$

It will be easy to confirm that the following equation

$$u = C_1 x + C_2 y + C_1{}^2 + C_2{}^2 \equiv \psi(x, y, C_1, C_2),$$

$$(C_1, C_2 : \text{arbitrary constants}) \tag{1.54}$$

satisfies the given PDE by a direct calculation, although the process of obtaining the above solution is left to a later subsection (§§1.4.4 **e**). The above solution includes two arbitrary constants, and hence is a *complete solution*. The latter gives a family of planes.

If these constants have a relation, *e.g.*, $C_2 = g(C_1)$, we have

$$u = C_1 x + g(C_1)y + C_1{}^2 + g(C_1)^2.$$

Using this equation, and the one partially differentiated with respect to C_1:

$$x + g'(C_1)y + 2C_1 + 2g(C_1)g'(C_1) = 0,$$

we can eliminate C_1 (or may regard them as a parametric representation in terms of C_1). Either of these equations gives a plane, and their intersection is a straight line. A change of C_1 gives the surface covered by a family of straight lines, which forms the integral surface. Since the present solution allows one arbitrary function, it is a *general solution* of Eq. (1.53).

Furthermore, if these constants are related, such as $C_2 = \sqrt{1 - C_1{}^2} \equiv g(C_1)$, then

$$u = C_1 x + \sqrt{1 - C_1{}^2}y + 1 \equiv \psi\left(x, y, C_1, \sqrt{1 - C_1{}^2}\right),$$

from which we have

$$\frac{\partial u}{\partial C_1} = 0 \quad \rightarrow \quad x - \frac{C_1 y}{\sqrt{1 - C_1{}^2}} = 0.$$

The elimination of C_1 from these two equations yields

$$(u - 1)^2 = x^2 + y^2,$$

which is an example of the *particular solution*.

Note that we have the following results from the complete solution (1.54):

$$\frac{\partial \psi}{\partial C_1} = x + 2C_1 = 0, \quad \frac{\partial \psi}{\partial C_2} = y + 2C_2 = 0,$$

so that $C_1 = -x/2$ and $C_2 = -y/2$. By substituting C_1 and C_2 for the complete solution (1.54), we have

$$u = -\frac{1}{4}(x^2 + y^2). \tag{1.55}$$

This equation fulfills Eq. (1.53), so that it is certainly a solution, but is by no means attainable with the choice of constants of the complete solution (1.54). This type of solution is an example of the *singular solution*.

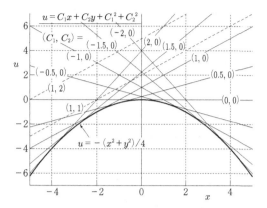

Fig. 1.5 Singular solution and the envelop surface (cross section at $y = 0$ is shown). Numerical values in the parentheses are (C_1, C_2).

Geometrically speaking, Eq. (1.55) shows a paraboloid of revolution. Considering the rotational symmetry, the cross-sectional view at $y = 0$ is shown in Fig. 1.5. The cross section of the complete solution (1.54) at $y = 0$ is a family of straight lines $u = C_1 x + C_1{}^2 + C_2{}^2$ in this figure, whereas the cross section of the paraboloid (1.55) is $u = -x^2/4$. In the former, every line with $C_2 = 0$ is tangent to the parabola $u = -x^2/4$, irrespective of the value of C_1, *i.e.*, the latter becomes an **envelop**. (For an "envelop", see also §§1.2.2 **b**.) Furthermore, every line with a varied C_2 is found in the region above that parabola. Viewed in a three-dimensional space, this shows that the quadric (1.55), *i.e.*, a paraboloid of revolution, becomes the envelop surface of the complete solution (1.54). ◁

1.4.2 *First-order PDEs with easily obtainable complete solutions*

In the following first-order PDEs, complete solutions are easily obtained. The method of the solution, however, will be given in §§1.4.4.

a. $f(p, q) = 0$ type

In this case, x, y, and u are not included in the PDE. In terms of two constants a and b that fulfill the relation $f(a, b) = 0$, the complete solution is given by

$$u = ax + by + C, \tag{1.56}$$

where C is a constant. Here, two arbitrary independent constants are a (or b) and C.

b. $f(x, p, q) = 0$ type

Since y is not included, we put $q \equiv \partial u / \partial y = b$, then we have

$$u = by + g(x), \qquad p \equiv \frac{\partial u}{\partial x} = g'(x),$$

and hence the PDE becomes $f(x, g'(x), b) = 0$. If we solve the latter, and $g'(x) \equiv h(x, b)$ is calculated, we can obtain the complete solution

$$u = \int h(x, b) \, dx + by + C, \qquad (b, \; C : \text{arbitrary constant}). \qquad (1.57)$$

Similarly, in the $f(y, p, q) = 0$ type, the complete solution is given by exchanging the roles of x and y.

c. $f(u, p, q) = 0$ type

We put $\xi = x + ay$, where a is an arbitrary constant, and seek a solution of the form $u = g(\xi)$. By a direct calculation, we have

$$p \equiv \frac{\partial u}{\partial x} = \frac{du}{d\xi} \frac{\partial \xi}{\partial x} = \frac{du}{d\xi}, \qquad q \equiv \frac{\partial u}{\partial y} = \frac{du}{d\xi} \frac{\partial \xi}{\partial y} = a \frac{du}{d\xi},$$

so that $u = g(\xi)$ fulfills the equation

$$f\left(u, \frac{du}{d\xi}, a \frac{du}{d\xi}\right) = 0.$$

If the above equation is solved for $du/d\xi$, and is given in the form $du/d\xi = h(u, a)$, then the latter is integrated by the method of separation of variables (known in the ODEs):

$$\xi = x + ay = \int \frac{du}{h(u, a)} + C, \qquad (1.58)$$

where a and C are two arbitrary constants. This is the complete solution.

d. $f(x, p) = g(y, q)$ type (separation of variables)

Taking into account that the variables are separated, we put $f(x, p) = g(y, q) = a$ (constant). If the respective equations are solved, and p and q are given

$$p \equiv \frac{\partial u}{\partial x} = \phi(x, a), \qquad q \equiv \frac{\partial u}{\partial y} = \psi(y, a),$$

then we obtain the complete solution

$$u = \int \phi(x, a) \, dx + \int \psi(y, a) \, dy + C, \qquad (1.59)$$

where a and C are two arbitrary constants.

1.4.3 *Integrable condition*

The total differential of the function $\Phi(x, y, u)$ of three independent variables x, y, u is given by

$$\mathrm{d}\Phi = \frac{\partial \Phi}{\partial x}\,\mathrm{d}x + \frac{\partial \Phi}{\partial y}\,\mathrm{d}y + \frac{\partial \Phi}{\partial u}\,\mathrm{d}u. \qquad (1.60)$$

Hence, if a given PDE is expressed in the form $\mathrm{d}\Phi = 0$, it is integrable and yields the solution $\Phi(x, y, u) = C$ (constant).

We shall now extend the above equation, and examine the condition in which the total differential equation with three independent variables x, y, u:

$$f(x, y, u)\,\mathrm{d}x + g(x, y, u)\,\mathrm{d}y + h(x, y, u)\,\mathrm{d}u = 0 \qquad (1.61)$$

is integrable.

Hinted by our knowledge on the integrable condition of the total differential equations with two independent variables,[15] we may expect the solution to be achieved in a similar manner. Namely, even if the given equation (1.61) cannot directly be expressed in the same form as Eq. (1.60), we may be able to introduce a certain function λ, such that the function Ψ fulfills

$$\lambda f = \frac{\partial \Psi}{\partial x}, \quad \lambda g = \frac{\partial \Psi}{\partial y}, \quad \lambda h = \frac{\partial \Psi}{\partial u}. \qquad (1.62)$$

---

[15]Consider the total differential equation with two independent variables x, y:

$$f(x, y)\mathrm{d}x + g(x, y)\,\mathrm{d}y = 0.$$

If a function Φ that fulfills $f = \partial\Phi/\partial x, g = \partial\Phi/\partial y$ is found, then the above equation becomes

$$\mathrm{d}\Phi \equiv \frac{\partial \Phi}{\partial x}\,\mathrm{d}x + \frac{\partial \Phi}{\partial y}\,\mathrm{d}y = 0,$$

so that the general solution $\Phi(x, y) = C$ is obtained, where C is an arbitrary constant. The condition in which the latter formulation is possible (an integrable condition) is

$$\frac{\partial f}{\partial y} = \frac{\partial g}{\partial x}\left(= \frac{\partial^2 \Phi}{\partial x \partial y}\right).$$

This type of total differential equation is called **exact** (or exact differential).

Even if a given equation itself is not exact, we may find a certain function $\lambda(x, y)$, such that

$$\lambda(x, y)f(x, y)\,\mathrm{d}x + \lambda(x, y)g(x, y)\,\mathrm{d}y = 0$$

is exact. Namely, if a function Ψ that satisfies the relation

$$\lambda f = \frac{\partial \Psi}{\partial x}, \qquad \lambda g = \frac{\partial \Psi}{\partial y},$$

is found, a general solution is similarly obtained. Such a function $\lambda(x, y)$ to facilitate the solution is called an **integrating factor**.

Then, Eq. (1.61) becomes

$$d\Psi = 0, \tag{1.63}$$

which yields the solution of Eq. (1.61)

$$\Psi(x, y, u) = C, \qquad (C : \text{arbitrary constant}). \tag{1.64}$$

A question remains under what conditions Eq. (1.61) has a solution of the form Eq. (1.63). To know this condition, we differentiate the first and second equations of (1.62) with respect to y and x, respectively:

$$\frac{\partial}{\partial y}\left(\frac{\partial \Psi}{\partial x}\right) = \frac{\partial \lambda}{\partial y} f + \lambda \frac{\partial f}{\partial y}, \qquad \frac{\partial}{\partial x}\left(\frac{\partial \Psi}{\partial y}\right) = \frac{\partial \lambda}{\partial x} g + \lambda \frac{\partial g}{\partial x}.$$

Since x and y are independent variables, two equations above must be the same, which yields

$$\lambda\left(\frac{\partial f}{\partial y} - \frac{\partial g}{\partial x}\right) = g \frac{\partial \lambda}{\partial x} - f \frac{\partial \lambda}{\partial y}.$$

Similarly, from the second and third equations of Eq. (1.62), we have

$$\lambda\left(\frac{\partial g}{\partial u} - \frac{\partial h}{\partial y}\right) = h \frac{\partial \lambda}{\partial y} - g \frac{\partial \lambda}{\partial u},$$

and from the third and first equations of Eq. (1.62), we have

$$\lambda\left(\frac{\partial h}{\partial x} - \frac{\partial f}{\partial u}\right) = f \frac{\partial \lambda}{\partial u} - h \frac{\partial \lambda}{\partial x}.$$

By multiplying these three equations by h, f, g, respectively, and summing up the respective sides, we obtain

$$f\left(\frac{\partial g}{\partial u} - \frac{\partial h}{\partial y}\right) + g\left(\frac{\partial h}{\partial x} - \frac{\partial f}{\partial u}\right) + h\left(\frac{\partial f}{\partial y} - \frac{\partial g}{\partial x}\right) = 0, \tag{1.65}$$

except for the trivial solution $\lambda = 0$. Equation (1.65) is the necessary condition for f, g, h to be written in the form of Eq. (1.63).

Conversely, if Eq. (1.65) holds, Eq. (1.64) gives the solution of Eq. (1.61), although the proof is omitted here. Equation (1.65) is called the **integrable condition**.

1.4.4 Lagrange–Charpit method

We apply the results obtained in the previous subsection to find the method for solving the first-order PDEs. The latter is generally given by

$$f(x, y, u, p, q) = 0, \qquad p = \frac{\partial u}{\partial x}, \qquad q = \frac{\partial u}{\partial y}, \tag{1.8}$$

where $u(x, y)$ is the unknown function of x and y. Since the total differential of u is given by

$$du = \frac{\partial u}{\partial x}\,dx + \frac{\partial u}{\partial y}\,dy = p\,dx + q\,dy,$$

or

$$p\,dx + q\,dy + (-1)\,du = 0, \tag{1.66}$$

the terms f, g, and h in Eq. (1.61) correspond to p, q, and -1, respectively, where we regard p and q as functions of x, y, u. If the present equation is integrable, Eq. (1.65) should be satisfied, so that we have

$$\frac{\partial q}{\partial x} - \frac{\partial p}{\partial y} + p\frac{\partial q}{\partial u} - q\frac{\partial p}{\partial u} = 0. \tag{1.67}$$

We now introduce an auxiliary function

$$\varphi(x, y, u, p, q) = C, \tag{1.68}$$

where C is an arbitrary constant. (If φ is determined, we can determine p and q as functions of x, y, u from Eq. (1.8) and Eq. (1.68).) To determine φ, we calculate the partial derivatives of Eqs. (1.8) and (1.68) with respect to x:

$$\frac{\partial f}{\partial x} + \frac{\partial f}{\partial p}\frac{\partial p}{\partial x} + \frac{\partial f}{\partial q}\frac{\partial q}{\partial x} = 0, \qquad \frac{\partial \varphi}{\partial x} + \frac{\partial \varphi}{\partial p}\frac{\partial p}{\partial x} + \frac{\partial \varphi}{\partial q}\frac{\partial q}{\partial x} = 0,$$

from which we obtain $\partial q/\partial x$ that appears in Eq. (1.67). Similarly, we obtain $\partial p/\partial y$ from the partial differentiation of Eqs. (1.8) and (1.68) with respect to y, as well as $\partial q/\partial u, \partial p/\partial u$ from the partial differentiation of these equations with respect to u. By substituting these results for the integrable condition (1.67), we obtain

$$\frac{\partial f}{\partial p}\frac{\partial \varphi}{\partial x} + \frac{\partial f}{\partial q}\frac{\partial \varphi}{\partial y} + \left(p\frac{\partial f}{\partial p} + q\frac{\partial f}{\partial q}\right)\frac{\partial \varphi}{\partial u}$$

$$-\left(\frac{\partial f}{\partial x} + p\frac{\partial f}{\partial u}\right)\frac{\partial \varphi}{\partial p} - \left(\frac{\partial f}{\partial y} + q\frac{\partial f}{\partial u}\right)\frac{\partial \varphi}{\partial q} = 0,$$

which is the Lagrange's PDE given by Eq. (1.41).

In accordance with the procedures shown in §§1.2.3, we solve the auxiliary equations of the form (1.42):

$$\frac{\mathrm{d}x}{\dfrac{\partial f}{\partial p}} = \frac{\mathrm{d}y}{\dfrac{\partial f}{\partial q}} = \frac{\mathrm{d}u}{p\dfrac{\partial f}{\partial p} + q\dfrac{\partial f}{\partial q}} = -\frac{\mathrm{d}p}{\dfrac{\partial f}{\partial x} + p\dfrac{\partial f}{\partial u}} = -\frac{\mathrm{d}q}{\dfrac{\partial f}{\partial y} + q\dfrac{\partial f}{\partial u}}, \qquad (1.69)$$

from which we can obtain the solution

$$p = \phi(x, y, u), \qquad q = \psi(x, y, u).$$

Then, Eq. (1.66):

$$\phi(x, y, u)\,\mathrm{d}x + \psi(x, y, u)\,\mathrm{d}y - \mathrm{d}u = 0 \qquad (1.70)$$

becomes integrable, so that the solution is obtained. This method of solution is called the **Lagrange–Charpit method**.[16] ◁

We consider again the examples given in §§1.4.2 **a–d**, by applying the Lagrange–Charpit method.

a. $f(p, q) = 0$ type

In this case, the auxiliary equation (1.69) is

$$\frac{\mathrm{d}x}{\dfrac{\partial f}{\partial p}} = \frac{\mathrm{d}y}{\dfrac{\partial f}{\partial q}} = \frac{\mathrm{d}u}{p\dfrac{\partial f}{\partial p} + q\dfrac{\partial f}{\partial q}} = \frac{\mathrm{d}p}{0} = \frac{\mathrm{d}q}{0},$$

so that we obtain $p = a$ and $q = b$, where a and b are constants.[17] From Eq. (1.70), we have

$$\mathrm{d}u = p\,\mathrm{d}x + q\,\mathrm{d}y = a\,\mathrm{d}x + b\,\mathrm{d}y,$$

which yields

$$u = a\,x + b\,y + C,$$

in agreement with Eq. (1.56). In the PDE of this type, $p = a$, $q = b$ (a, b being arbitrary constants) give the solution. However, they are not independent to each other, and need to fulfill the relation $f(a, b) = 0$, so that the two arbitrary constants are a (or b) and C.

Example 1.14. Obtain the complete solution of $p^2 + q^2 = 1$.

---

[16]We need not solve the auxiliary equation (1.69) in general. If we obtain a solution including p and/or q, then it is sufficient to substitute them for Eq. (1.66).

[17]The meaning of "0" in the denominator, as appears in the expression $\mathrm{d}p/0$ or $\mathrm{d}q/0$, is the same as that noted previously in footnote 12 in §§1.2.3.

According to the previously-mentioned procedures, the complete solution is given by putting $p = a$, $q = b$, which yields $u = ax + by + C$. Here, a, b, C are "arbitrary" constants, although the former two must satisfy the condition $a^2 + b^2 = 1$. An alternative description is

$$u = ax \pm \sqrt{1 - a^2}\,y + C, \qquad (a,\, C : \text{arbitrary constants}),$$

or

$$u = x \cos\alpha + y \sin\alpha + C, \qquad (\alpha,\, C : \text{arbitrary constants}),$$

where we put $a = \cos\alpha$ and $b = \sin\alpha$. ◁

b. $f(x, p, q) = 0$ type

In this case, the auxiliary equation (1.69) becomes

$$\frac{\mathrm{d}x}{\dfrac{\partial f}{\partial p}} = \frac{\mathrm{d}y}{\dfrac{\partial f}{\partial q}} = \frac{\mathrm{d}u}{p\dfrac{\partial f}{\partial p} + q\dfrac{\partial f}{\partial q}} = -\frac{\mathrm{d}p}{\dfrac{\partial f}{\partial x}} = -\frac{\mathrm{d}q}{0},$$

so that we obtain $q = b$ (constant), and hence $f(x, p, b) = 0$. If we solve the latter equation, and the solution $p = \omega(x, b)$ is obtained, Eq. (1.70) yields

$$\mathrm{d}u = \omega(x, b)\,\mathrm{d}x + b\,\mathrm{d}y,$$

which is integrated to give the complete solution

$$u = \int \omega(x, b)\,\mathrm{d}x + by + C, \qquad (b,\, C : \text{arbitrary constants}),$$

in agreement with Eq. (1.57). In short, it is sufficient to put $q = b$ (constant) for this type of PDE.

Example 1.15. Obtain the complete solution of $xp - q^2 = 0$.

By putting $q = b$ in the given equation, we have $p = b^2/x$, hence we have

$$\mathrm{d}u \equiv p\,\mathrm{d}x + q\,\mathrm{d}y = b^2\,\frac{\mathrm{d}x}{x} + b\,\mathrm{d}y.$$

By integration, we obtain the complete solution

$$u = b^2 \log x + b\,y + C, \qquad (b,\, C : \text{arbitrary constants}).$$

◁

c. $f(u, p, q) = 0$ type

In this case, the auxiliary equation (1.69) becomes

$$\frac{dx}{\dfrac{\partial f}{\partial p}} = \frac{dy}{\dfrac{\partial f}{\partial q}} = \frac{du}{p\dfrac{\partial f}{\partial p} + q\dfrac{\partial f}{\partial q}} = -\frac{dp}{p\dfrac{\partial f}{\partial u}} = -\frac{dq}{q\dfrac{\partial f}{\partial u}}.$$

Using the last two terms, we have

$$\frac{dp}{p} = \frac{dq}{q},$$

and hence $q = ap$ (with a being an arbitrary constant). By substitution of q for the given PDE, we have $f(u, p, ap) = 0$. If the latter is solved, and $p = \omega(u, a)$ is obtained, Eq. (1.70) becomes $du = \omega(u, a)(dx + a\, dy)$. Following the method of separation of variables, we obtain the complete solution:

$$x + ay = \int \frac{du}{\omega(u, a)} + C, \qquad (a, C : \text{arbitrary constants}),$$

in agreement with Eq. (1.58). In short, it is sufficient to put $q = ap$ for this type of PDE.

Example 1.16. Obtain the complete solution of $upq - p - q = 0$.

In accordance with the previously-mentioned procedures, we put $q = ap$, and find the solutions p and q:

$$uap^2 - p - ap = 0 \qquad \rightarrow \qquad p = q = 0, \quad \text{or} \quad p = \frac{1+a}{au}, \quad q = \frac{1+a}{u}.$$

The former yields $u = $ constant, which is a trivial solution. Meanwhile, the latter yields

$$du \equiv p\, dx + q\, dy = \frac{1+a}{au}(dx + a\, dy),$$

or

$$\frac{1}{2} du^2 = \frac{1+a}{a}(dx + a\, dy),$$

which is integrated to yield the complete solution

$$u^2 = \frac{2(1+a)}{a}(x + ay) + C, \qquad (a, C : \text{arbitrary constants}).$$

◁

d. $f(x, p) = g(y, q)$ type (separation of variables)

In this case, the auxiliary equation (1.69) is considered for $f(x, p) - g(y, q)$ in place of f:

$$\frac{\mathrm{d}x}{\dfrac{\partial f}{\partial p}} = -\frac{\mathrm{d}y}{\dfrac{\partial g}{\partial q}} = \frac{\mathrm{d}u}{p\dfrac{\partial f}{\partial p} - q\dfrac{\partial g}{\partial q}} = -\frac{\mathrm{d}p}{\dfrac{\partial f}{\partial x}} = \frac{\mathrm{d}q}{\dfrac{\partial g}{\partial y}}.$$

Equating the first and fourth terms, and the second and fifth terms, we have

$$\frac{\partial f}{\partial x}\mathrm{d}x + \frac{\partial f}{\partial p}\mathrm{d}p = 0 \ (= \mathrm{d}f), \qquad \frac{\partial g}{\partial y}\mathrm{d}y + \frac{\partial g}{\partial q}\mathrm{d}q = 0 \ (= \mathrm{d}g),$$

and hence, $f(x, p) = a(\text{constant}) = g(y, q)$. From these equations, we calculate $p = \omega(x, a)$, $q = \chi(y, a)$, and substitute them for Eq. (1.70), which yields

$$\mathrm{d}u = \omega(x, a)\,\mathrm{d}x + \chi(y, a)\,\mathrm{d}y.$$

The integration of the latter gives the complete solution

$$u = \int \omega(x, a)\,\mathrm{d}x + \int \chi(y, a)\,\mathrm{d}y + C, \qquad (a, C : \text{arbitrary constants}),$$

in agreement with Eq. (1.59). In short, it is sufficient for this type of PDE to equate the respective sides to a constant a, and solve p and q, separately.

Example 1.17. Obtain the complete solution of $p + q = x^2 + y^2$.

It is easily found that the variables are separated by rewriting the given PDE to $p - x^2 = -q + y^2$. By putting the respective sides equal to a constant a, we have

$$p = x^2 + a, \qquad q = y^2 - a,$$

from which we have

$$\mathrm{d}u \equiv p\,\mathrm{d}x + q\,\mathrm{d}y = (x^2 + a)\,\mathrm{d}x + (y^2 - a)\,\mathrm{d}y.$$

Integration of the latter yields the complete solution

$$u = \frac{1}{3}(x^3 + y^3) + a(x - y) + C, \quad (a, C : \text{arbitrary constants}).$$

$\triangleleft$

e. Clairaut's equation

The type of the ODEs, $y = xp + f(p)$ is known as the Clairaut's equation. The latter is extended to the first-order PDE, and is expressed by the following type:

$$u = xp + yq + f(p, q).$$

The latter is also called the **Clairaut's equation**. In this case, the auxiliary equation (1.69) becomes the following (we may replace f with $xp + yq + f(p, q) - u$ and calculate the auxiliary equation):

$$\frac{\mathrm{d}x}{x + \dfrac{\partial f}{\partial p}} = \frac{\mathrm{d}y}{y + \dfrac{\partial f}{\partial q}} = \frac{\mathrm{d}u}{xp + yq + p\dfrac{\partial f}{\partial p} + q\dfrac{\partial f}{\partial q}} = -\frac{\mathrm{d}p}{0} = -\frac{\mathrm{d}q}{0},$$

from which we obtain $p = a$ and $q = b$ (a, b being arbitrary constants), and hence the complete solution

$$u = ax + by + f(a, b).$$

This solution gives a family of planes, and has a singular solution (envelop) in general. In short, it is sufficient for this type of PDE to put $p = a$ and $q = b$ (a, b: constants).

Example 1.18. Obtain the complete solution of

$$u = xp + yq + p^2 + q^2,$$

which is the Clairaut-type PDE that has already dealt with by Eq. (1.53).

In accordance with the above-mentioned procedures, we put $p = a$ and $q = b$, which yields the complete solution

$$u = ax + by + a^2 + b^2,$$

where a and b are arbitrary constants (Eq. (1.54)). The latter gives a family of planes, and has a singular solution (envelop). See Example 1.13.   ◁

1.5   Hamilton–Jacobi Theory

1.5.1   *Extremum problems and the Euler–Lagrange equation*

We first consider the extremum problems of a function $y = f(x)$. As shown in Fig. 1.6(a), we assume that a given smooth function f takes a **local extremum** at some point $x = a$. We calculate the difference δf

between the value of the function evaluated at a point $x = a$ and that of the neighboring point at an infinitesimal distance ϵ apart:

$$\delta f \equiv f(a + \epsilon) - f(a) = \epsilon f'(a) + O(\epsilon^2). \qquad (1.71)$$

If $x = a$ is a **stationary point** of the function f, it implies that $\delta f = o(\epsilon)$, so that $f'(a) = 0$ is the condition that characterizes the **extremum**. Whether the latter point is a **maximum** or a **minimum**, however, requires a further examination of higher order variation (*i.e.*, second derivative).

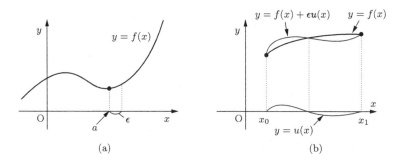

(a)   (b)

Fig. 1.6   Extremum of (a) the function, and (b) the functional.

We now consider the extremum problems of a functional. In general, a function relates the "numerical value" to the "numerical value", whereas a **functional** relates the "function" to the "function". In practice, we sometimes meet a problem, in which the function to be minimized (or maximized) is included in the integrand. Here, the range of integration of the independent variable as well as the boundary conditions are given, but the exact form of the dependent variable is not known. For example, when we want to know the area S between some smooth function $y = f(x)$ and $y = 0$ in the interval $[x_0, x_1]$, the numerical value S is calculable by integration if the function $f(x)$ is given. In this case, a certain "numerical value" is assigned to a given "function". We call such a type of functional an **integral functional**.

In the following, we confine our attention to the integral J of a smooth function F over the specified interval $[x_0, x_1]$, where F includes the unknown function y and its first derivative $y' \equiv v$, so that $F = F(y, v, x)$ and

$$J[y] \equiv \int_{x_0}^{x_1} F(y, v, x)\, dx, \quad \text{where} \quad y(x_0) = y_0, \quad y(x_1) = y_1. \qquad (1.72)$$

Our problem is to find the function y, which makes J extremum subject to the constraints on the boundary conditions $y(x_0) = y_0$ and $y(x_1) = y_1$.

Similar to what we did for finding the extremum of a function, we calculate the increment of the integral δJ (called **variation**) with the change of y. To be more precise, we assume that J takes an extremum when y is equal to $f(x)$ (existence of such a function is assumed), and calculate the difference owing to the change of the function y by an infinitesimal amount $\epsilon u(x)$. Here, the function $u(x)$ is assumed to be arbitrary except that it fulfills the constraints $u(x_0) = u(x_1) = 0$ in accordance with the boundary condition (see Fig. 1.6(b)). Corresponding to Eq. (1.71), we have

$$\delta J = \int_{x_0}^{x_1} [F(y + \epsilon u, v + \epsilon u', x) - F(y, v, x)]\, \mathrm{d}x$$

$$= \int_{x_0}^{x_1} \left[\epsilon u \frac{\partial F}{\partial y} + \epsilon u' \frac{\partial F}{\partial v} \right] \mathrm{d}x + O(\epsilon^2)$$

$$= \epsilon \int_{x_0}^{x_1} u \left[\frac{\partial F}{\partial y} - \frac{\mathrm{d}}{\mathrm{d}x}\left(\frac{\partial F}{\partial v} \right) \right] \mathrm{d}x + O(\epsilon^2),$$

where we have integrated by parts in the second integral, and the boundary conditions on u are considered. By our assumption that $y = f(x)$ gives the stationary value to J, we have $\delta J = 0$ at $O(\epsilon)$:

$$\int_{x_0}^{x_1} u(x) \left[\frac{\partial F}{\partial y} - \frac{\mathrm{d}}{\mathrm{d}x}\left(\frac{\partial F}{\partial v} \right) \right] \mathrm{d}x = 0.$$

Since $u(x)$ is an arbitrary function except for the restriction imposed at $x = x_0$ and x_1, the above equation holds only if

$$\frac{\partial F}{\partial y} - \frac{\mathrm{d}}{\mathrm{d}x}\left(\frac{\partial F}{\partial v} \right) = 0, \tag{1.73}$$

which is called the **Euler–Lagrange equation**. Whether the function obtained gives the maximal or minimal values needs the second order variation. ◁

Note that the Euler–Lagrange equation (1.73) does not include the function F itself. Hence, the addition of the total derivative of an arbitrary function with respect to x to F does not change the equation. Namely, if we add a total derivative $\mathrm{d}\tilde{f}(y, v, x)/\mathrm{d}x$ to the integrand of Eq. (1.72), then we have

$$\tilde{J}[y] \equiv \int_{x_0}^{x_1} \left(F(y, v, x) + \frac{\mathrm{d}\tilde{f}}{\mathrm{d}x} \right) \mathrm{d}x = \int_{x_0}^{x_1} F(y, v, x)\, \mathrm{d}x + \left[\tilde{f}(y, v, x) \right]_{x_0}^{x_1}.$$

$$\tag{1.74}$$

Since the terms including $\tilde{f}$ are fixed at both ends of the interval x_0 and x_1, the variation of $\tilde{f}$ vanishes, *i.e.*, $\delta\tilde{J} = \delta J$, so that it has no effect on the final result. This feature is called the **"indeterminacy"** of the integrand in variation.

The above-mentioned method is known as the **calculus of variations**.

Example 1.19 (Curve of steepest descent[18]). Obtain the shape of the slope (or roller-coaster), along which an object slides down between the two points in the vertical plane with the shortest time.

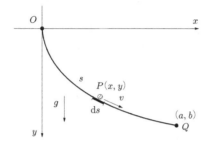

Fig. 1.7   Curve of steepest descent (Brachistochrone).

To formulate this problem more precisely, we make some assumptions. As shown in Fig. 1.7, we choose the x axis horizontally, and y axis in the vertical direction downward. We choose the starting position at the origin O, and the goal at $Q(a, b)$. We denote the mass of the object by m, and the gravitational acceleration by g, which is constant and is directed vertically downward. We also denote the speed of the object at a point $P(x, y)$ along the slope by v. If we assume that the size of the object, as well as the frictional force between the object and the slope, are negligible, we can apply the conservation law of the total mechanical energy, well-known in physics, so that the relation "$\frac{1}{2}mv^2 + mg(-y) = $ constant" holds. Then, the object initially at rest at point O attains a speed $v = \sqrt{2gy}$ at point P. If we denote the length along the slope from the origin by s, then the line segment $\mathrm{d}s$ in the neighborhood of the point P is given by $\mathrm{d}s = \sqrt{(\mathrm{d}x)^2 + (\mathrm{d}y)^2} = \sqrt{1 + y'^2}\,\mathrm{d}x$, and the time $\mathrm{d}t$ required for the object to pass through this segment is given by $\mathrm{d}t = \mathrm{d}s/v$. Therefore, the time T necessary for this object to move from point O to point Q is given

---

[18]or "Brachistochrone" in German.

by

$$T = \int_O^Q \frac{\mathrm{d}s}{v} = \int_0^a \frac{\sqrt{1+y'^2}}{\sqrt{2gy}}\,\mathrm{d}x = \frac{1}{\sqrt{2g}}\int_0^a F(y,y')\,\mathrm{d}x,$$

$$\text{where} \quad F(y,y') = \sqrt{\frac{1+y'^2}{y}}.$$

The above equation is of the same form as Eq. (1.72), so that y should satisfy Eq. (1.73) to reach the goal with shortest time (note that, from the viewpoint of physics, it is trivial that such an extremum corresponds to a minimum and not to a maximum).

In general, if the integrand of Eq. (1.72) depends only on y and $y'(= v)$, *i.e.*, $F = F(y,y')$, Eq. (1.73) can be further integrated. Namely, by multiplying y' to both sides of Eq. (1.73), the latter is rewritten as

$$y'\left[\frac{\partial F}{\partial y} - \frac{\mathrm{d}}{\mathrm{d}x}\left(\frac{\partial F}{\partial y'}\right)\right] = \frac{\mathrm{d}}{\mathrm{d}x}\left(F - y'\frac{\partial F}{\partial y'}\right) = 0,$$

so that it is integrated[19] to yield

$$F - y'\frac{\partial F}{\partial y'} = C, \qquad (1.75)$$

where C is an arbitrary constant.

In our present example, the substitution of $F(y,y') = \sqrt{(1+y'^2)/y}$ for Eq. (1.75) yields

$$\sqrt{\frac{1+y'^2}{y}} - \frac{y'^2}{\sqrt{y(1+y'^2)}} = C, \qquad \rightarrow \qquad \frac{\mathrm{d}y}{\mathrm{d}x} = \sqrt{\frac{k-y}{y}}, \qquad (1.76)$$

where $k \equiv 1/C^2$ is an arbitrary positive constant. In the above calculation, we take account of $\mathrm{d}y/\mathrm{d}x \geq 0$ from the viewpoint of physics. Equation (1.76) is an ODE of a separable-variable type, so that it is easily

---

[19]Subtract the following:

$$\frac{\mathrm{d}}{\mathrm{d}x}F(y,y') = \frac{\partial F}{\partial y}\frac{\mathrm{d}y}{\mathrm{d}x} + \frac{\partial F}{\partial y'}\frac{\mathrm{d}y'}{\mathrm{d}x} = y'\frac{\partial F}{\partial y} + \frac{\mathrm{d}y'}{\mathrm{d}x}\frac{\partial F}{\partial y'},$$

$$\frac{\mathrm{d}}{\mathrm{d}x}\left(y'\frac{\partial F}{\partial y'}\right) = \frac{\mathrm{d}y'}{\mathrm{d}x}\frac{\partial F}{\partial y'} + y'\frac{\mathrm{d}}{\mathrm{d}x}\left(\frac{\partial F}{\partial y'}\right).$$

integrated[20]

$$x = \frac{k}{2}(\theta - \sin\theta), \qquad y = \frac{k}{2}(1 - \cos\theta). \tag{1.77}$$

The curve Eq. (1.77) is well-known as a cycloid. A cycloid is the curve traced by a marked point on a circle when the latter rolls along the horizontal plane without slipping (see Fig. 1.8(a)). In particular, the point with $\theta = 2\pi$ is the position where the marker particle passes through the highest y position (the lowest position in Fig. 1.7 and Fig. 1.8(b)) on the orbit and rolls down to reach the same height again. The horizontal distance l traversed is $2\pi a$, and the time T necessary to reach that position is

$$T = \sqrt{\frac{2\pi l}{g}} = 2\pi\sqrt{\frac{a}{g}}.$$

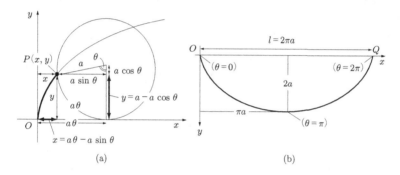

(a)   (b)

Fig. 1.8   Cycloid.

For example, if we connect the two points of the same height with horizontal distance l apart by the roller coaster of a cycloidal shape (see Fig. 1.8(b), where y axis is directed vertically downward), the travel time T is 8.01

---

[20]From Eq. (1.76), we separate the variables x and y, and integrate the respective sides. In terms of the parameter θ that fulfills $y = k\sin^2(\theta/2) = (k/2)(1 - \cos\theta)$, we have

$$x = \int^y \sqrt{\frac{y}{k-y}}\, dy = k \int^\theta \sin^2\frac{\theta}{2}\, d\theta = \frac{k}{2}(\theta - \sin\theta) + C_1,$$

where C_1 is an arbitrary constant. By the assumption that the starting point is the origin O ($x = y = 0$), at which we choose $\theta = 0$, we have $C_1 = 0$. Consequently, we have

$$y = \frac{k}{2}(1 - \cos\theta), \qquad x = \frac{k}{2}(\theta - \sin\theta).$$

The remaining constant k is determined so as to pass the point Q.

seconds for a distance of $l = 100$ m, 25.3 seconds for $l = 1$ km, 4 minutes
13 seconds for $l = 100$ km, 13 minutes 21 seconds for $l = 1000$ km, etc.

Further details on the calculus of variations are given in *e.g.*, [Morse
and Feshbach (1953); Byron and Fuller (1969); Arfken (1985)], *etc.*

1.5.2 *Lagrange's equation and canonical equation*

Many problems in classical mechanics are reduced to the extremum problems of the integral functional J ([Morse and Feshbach (1953); Byron and
Fuller (1969); Arfken (1985)] *etc.*), where integrand L of the latter is a
function of the coordinates x and velocity v $(= \mathrm{d}x/\mathrm{d}t)$:

$$J = \int_{t_0}^{t_1} L(x, v, t)\, \mathrm{d}t. \tag{1.78}$$

In classical mechanics, the integrand L is called the **Lagrangian**. We
calculate the variation δJ under the constraints $x(t_0) = x_0$, $x(t_1) = x_1$
in the same manner as before, which yields the **Lagrange's equation of
motion**:

$$\frac{\mathrm{d}}{\mathrm{d}t}\left(\frac{\partial L}{\partial v}\right) - \frac{\partial L}{\partial x} = 0. \tag{1.79}$$

If we define

$$p = \frac{\partial L}{\partial v}, \tag{1.80}$$

and find the expression v in terms of p, we can transform the set of variables
(x, v, t) with $L(x, v, t)$ to (x, p, t) with $H(x, p, t) \equiv p\, v - L$.[21] In classical
mechanics, H has the meaning of energy, and the same quantity expressed
by the coordinate and momentum is called the **Hamiltonian**. The functions H and L are reciprocal, and the variable p is transformed to v by the
transformation $L = p\, v - H$, so that we have

$$v = \frac{\partial H}{\partial p}\left(= \frac{\mathrm{d}x}{\mathrm{d}t}\right), \tag{1.81}$$

and Eq. (1.79) becomes

$$\frac{\mathrm{d}p}{\mathrm{d}t} = \frac{\partial L}{\partial x} = -\frac{\partial H}{\partial x}. \tag{1.82}$$

---

[21]The transformation of a set of conjugate variables is generally called the **Legendre
transformation**. In the present example, (v, p) is a set of conjugate variables, so that
the transformation of independent variables x, v, t in one function L to x, p, t in the other
function H is the Legendre transformation.

Consequently, we have a pair of equations (1.81) and (1.82):

$$\frac{\mathrm{d}x}{\mathrm{d}t} = \frac{\partial H}{\partial p}, \qquad \frac{\mathrm{d}p}{\mathrm{d}t} = -\frac{\partial H}{\partial x}, \tag{1.83}$$

which is called the **canonical equation**, or the **Hamilton's equation**. ◁

Note that the previously-mentioned relations are extended to the quantities that depend on many more variables. If we denote the independent coordinates and velocities by x_i and v_i ($i = 1, 2, \ldots, n$), respectively, so that the Lagrangian is given by $L = L(x_1, \ldots, x_n, v_1, \ldots, v_n, t)$, then Eq. (1.79) and Eq. (1.80) become

$$\frac{\mathrm{d}}{\mathrm{d}t}\left(\frac{\partial L}{\partial v_i}\right) - \frac{\partial L}{\partial x_i} = 0, \qquad (i = 1, 2, \ldots, n), \tag{1.84}$$

and

$$p_i = \frac{\partial L}{\partial v_i}, \qquad (i = 1, 2, \ldots, n). \tag{1.85}$$

By the Legendre transformation $L = \sum\limits_{i=1}^{n} p_i v_i - H$, the Hamiltonian becomes $H = H(x_1, \ldots, x_n, p_1, \ldots, p_n, t)$, so that the canonical equations are given by

$$\frac{\mathrm{d}x_i}{\mathrm{d}t} = \frac{\partial H}{\partial p_i}, \qquad \frac{\mathrm{d}p_i}{\mathrm{d}t} = -\frac{\partial H}{\partial x_i}, \qquad (i = 1, 2, \ldots, n). \tag{1.86}$$

The Lagrange's equations of motion rewrite the Newton's equations of motion in a general manner as a time derivative of the momenta and derivatives of the scalar function of the coordinates and velocities. The resulting equations are a set of the same number of second-order ODEs as the Newton's. On the other hand, the Hamilton's equations are a set of the double number of first-order ODEs, and describe the motion as an orbit of the point in the phase space of the coordinates and momenta.

1.5.3 *Canonical transform*

We consider the transformation of a set of coordinates and momenta x_i, p_i ($n = 1, \ldots, n$) to a new set of variables Q_i, P_i ($n = 1, \ldots, n$). We assume that $H(x_i, p_i, t)$ is transformed to $K(Q_i, P_i, t)$. If the following relation:

$$\frac{\mathrm{d}Q_i}{\mathrm{d}t} = \frac{\partial K}{\partial P_i}, \qquad \frac{\mathrm{d}P_i}{\mathrm{d}t} = -\frac{\partial K}{\partial Q_i} \tag{1.87}$$

holds by this transformation, we call the latter a **canonical transform**.

For a transform to be canonical, the variation of the Lagrangian:

$$\delta \left(\int_{t_0}^{t_1} L \, dt \right) = \delta \left(\int_{t_0}^{t_1} \left[\sum_{i=1}^{n} \dot{x}_i p_i - H(x_i, p_i, t) \right] dt \right)$$

in terms of the new variables must also be zero. For this purpose, we require that the integrands expressed by old and new variables are equal to each other except for the total derivative of an arbitrary function (as mentioned as indeterminacy in Eq. (1.74)), which yields

$$\dot{x}_i p_i - H(x_i, p_i, t) = \dot{Q}_i P_i - K(Q_i, P_i, t) + \frac{d}{dt} F, \qquad (1.88)$$

where, for brevity, we have denoted the differentiation with respect to time by over-dot. Note also that, here and hereafter, we adopt the "summation convention" introduced by Einstein, in which the repeated suffixes of the same number, such as $a_i b_i$, imply that the summation is taken over that suffix, and that the summation symbol Σ is omitted. The function F is called the **generating function**.   ◁

There are four types of generating functions, with the use of a set $(x_i, Q_i), (x_i, P_i), (p_i, Q_i), (p_i, P_i)$. These are examples of the Legendre transformation, and are transformed to each other by a change of variables in the generating functions.

a. $F_1(x_i, Q_i, t)$ type

In this case, a direct calculation of Eq. (1.88) using $F = F_1(x_i, Q_i, t)$ shows that

$$\dot{Q}_i P_i - K(Q_i, P_i, t) + \frac{d}{dt} F_1(x_i, Q_i, t) = \dot{Q}_i P_i - K + \frac{\partial F_1}{\partial t} + \frac{\partial F_1}{\partial x_i} \dot{x}_i + \frac{\partial F_1}{\partial Q_i} \dot{Q}_i$$

$$= \dot{x}_i p_i - H,$$

so that the required conditions are

$$P_i = -\frac{\partial F_1}{\partial Q_i}, \quad p_i = \frac{\partial F_1}{\partial x_i}, \quad K = H + \frac{\partial F_1}{\partial t}. \qquad (1.89)$$

b. $F_2(x_i, P_i, t)$ type

In order to change the variables from Q_i to P_i in the previous case **a**, we put $F = F_2(x_i, P_i, t) - P_i Q_i,$[22] which yields

$$p_i = \frac{\partial F_2}{\partial x_i}, \quad Q_i = \frac{\partial F_2}{\partial P_i}, \quad K = H + \frac{\partial F_2}{\partial t}. \tag{1.90}$$

c. $F_3(Q_i, p_i, t)$ type

Similarly, a change of x_i to p_i from the case **a** is made by using $F = F_3(Q_i, p_i, t) + p_i x_i$:

$$P_i = -\frac{\partial F_3}{\partial Q_i}, \quad x_i = -\frac{\partial F_3}{\partial p_i}, \quad K = H + \frac{\partial F_3}{\partial t}. \tag{1.91}$$

d. $F_4(p_i, P_i, t)$ type

A similar calculation using $F = F_4(p_i, P_i, t) - P_i Q_i + p_i x_i$ yields

$$Q_i = \frac{\partial F_4}{\partial P_i}, \quad x_i = -\frac{\partial F_4}{\partial p_i}, \quad K = H + \frac{\partial F_4}{\partial t}. \tag{1.92}$$

◁

Example 1.20. If we use $F_2 = x_i P_i$, we have $p_i = P_i$, $Q_i = x_i$, which is the identity transformation. ◁

Example 1.21. If we use $F_1 = x_i Q_i$, we have $p_i = Q_i$, $P_i = -x_i$, in which the coordinates and momenta are exchanged (except for the minus sign). As this example shows, the distinction of the coordinates and momenta becomes meaningless after the canonical transform, and both are treated as general variables. ◁

1.5.4 *Hamilton–Jacobi equation*

If an appropriate generating function F is found, and the associated canonical transform $(x_i, p_i) \to (Q_i, P_i)$ yields a new Hamiltonian K that is identically equal to zero, then Eq. (1.87) becomes

$$\dot{Q}_i = \frac{\partial K}{\partial P_i} = 0, \quad \dot{P}_i = -\frac{\partial K}{\partial Q_i} = 0, \quad (i = 1, \dots, n).$$

---

[22]The choice of variables can also be shown by a direct calculation. The use of $F = F_2(x_i, P_i, t)$ yields

$$\dot{x}_i p_i - H = \dot{Q}_i P_i - K(Q_i, P_i, t) + \frac{\partial F_2}{\partial t} + \frac{\partial F_2}{\partial x_i} \dot{x}_i + \frac{\partial F_2}{\partial P_i} \dot{P}_i,$$

which has no terms comparable to $\dot{P}_i$. In this case, however, we may rewrite the first term of the r.h.s. as $\dot{Q}_i P_i = \frac{d}{dt}(Q_i P_i) - Q_i \dot{P}_i$ and include the total derivative $\frac{d}{dt}(Q_i P_i)$ into the generating function.

The latter equations imply

$$P_i = \alpha_i \text{ (constant)}, \qquad Q_i = \beta_i \text{ (constant)}, \qquad (i = 1, \ldots, n).$$

If they are solved, and yields

$$x_i = x_i(\alpha_1, \ldots, \alpha_n, \beta_1, \ldots, \beta_n, t), \qquad p_i = p_i(\alpha_1, \ldots, \alpha_n, \beta_1, \ldots, \beta_n, t),$$

then the solution is obtained. Thus, our task is to find the appropriate generating function F mentioned above. To do so, we adopt the **b-type** generating function of the previous subsection (§§1.5.3b), *i.e.*,

$$Q_i = \frac{\partial F}{\partial P_i}, \qquad p_i = \frac{\partial F}{\partial x_i},$$

and

$$H\left(x_1, \ldots, x_n, \frac{\partial F}{\partial x_1}, \cdots, \frac{\partial F}{\partial x_n}, t\right) + \frac{\partial F}{\partial t} = 0. \tag{1.93}$$

Equation (1.93) is called the **Hamilton–Jacobi equation**. The latter is a first-order PDE for F with $(n+1)$ independent variables $x_1, \ldots, x_n, t$.

1.6   Integral of the Hamilton–Jacobi equation

In many problems in classical mechanics, we need to know the complete solution (the solution including the same number of independent constants as the number of independent variables) of the Hamilton–Jacobi equation rather than a general solution. Thus, it is important to find the solution of Eq. (1.93) with $(n+1)$ constants. Since Eq. (1.93) includes only partial derivatives of F, the addition of a constant gives no effect at all. Taking the latter into account, we put $F = W + \alpha_{n+1}$ (constant), then W fulfills

$$H\left(x_1, \ldots, x_n, \frac{\partial W}{\partial x_1}, \cdots, \frac{\partial W}{\partial x_n}, t\right) + \frac{\partial W}{\partial t} = 0, \tag{1.94}$$

where α_{n+1} is an arbitrary constant but does not appear explicitly in the resulting mechanics. The function W is called the **Hamilton's principal function**, which reduces the unknown constants by one. Our next task is to find the solution including n constants α_i $(i = 1, \ldots, n)$:

$$W = W(x_1, \ldots, x_n, \alpha_1, \ldots, \alpha_n, t). \tag{1.95}$$

Once these constants α_i are known, we choose them as the new momenta:

$$P_i = \alpha_i, \qquad (i = 1, \ldots, n), \tag{1.96}$$

and put

$$Q_i = \frac{\partial W}{\partial P_i} = \frac{\partial}{\partial \alpha_i} W(\boldsymbol{x}, \boldsymbol{\alpha}, t) = \beta_i, \qquad (i = 1, \ldots, n), \tag{1.97}$$

as new coordinates. Here, we have denoted $\boldsymbol{x} = (x_1, \ldots, x_n)$, $\boldsymbol{\alpha} = (\alpha_1, \ldots, \alpha_n)$, $\boldsymbol{\beta} = (\beta_1, \ldots, \beta_n)$, for brevity. By solving Eq. (1.97) for x_i, we have

$$x_i = \phi_i(\boldsymbol{\alpha}, \, \boldsymbol{\beta}, t), \qquad (i = 1, \ldots, n), \tag{1.98}$$

then we have

$$p_i = \frac{\partial}{\partial x_i} W\big(\boldsymbol{x}(\boldsymbol{\alpha}, \, \boldsymbol{\beta}, t), \boldsymbol{\alpha}, t\big) = \psi_i(\boldsymbol{\alpha} \, \boldsymbol{\beta}, t), \qquad (i = 1, \ldots, n). \tag{1.99}$$

Hence we obtain the solution including $2n$ constants. ◁

Our remaining task is to find W. To do so, we follow the general method of solution for the first-order quasi-linear PDE. Namely, we put

$$\frac{\partial W}{\partial x_1} = p_1, \quad \cdots, \quad \frac{\partial W}{\partial x_n} = p_n, \quad \frac{\partial W}{\partial t} = p, \tag{1.100}$$

into Eq. (1.94):

$$H(x_1, \ldots, x_n, p_1, \ldots, p_n, t) + p = 0, \tag{1.101}$$

and apply the Lagrange–Charpit method. In particular, Eq. (1.101) does not explicitly include the dependent variable W, so that the auxiliary equations (1.69) become

$$\frac{\mathrm{d}x_1}{\dfrac{\partial H}{\partial p_1}} = \cdots = \frac{\mathrm{d}x_n}{\dfrac{\partial H}{\partial p_n}} = \frac{\mathrm{d}t}{1} = -\frac{\mathrm{d}p_1}{\dfrac{\partial H}{\partial x_1}} = \cdots = -\frac{\mathrm{d}p_n}{\dfrac{\partial H}{\partial x_n}}, \tag{1.102}$$

which are solved for the given problems. The combination of the term $\mathrm{d}t/1$ with other terms reproduces the Hamilton's equation Eq. (1.86). ◁

Example 1.22 (Free particle). As a simple example, we consider a free particle. Here, a "free particle" refers to a particle that is not bounded by an external force, so that the total mechanical energy is solely given by the kinetic energy.

We denote the position of the particle in a three-dimensional space by (x_1, x_2, x_3), its velocity by (v_1, v_2, v_3), and momenta by (p_1, p_2, p_3). In accordance with the notation in this subsection, the coordinates (x_1, x_2, x_3) are used in place of (x, y, z), and the momenta are given by $p_i = m v_i$ ($i = 1, 2, 3$), where m is the mass of the particle.

In this problem, the Hamiltonian is given by[23]

$$H = \frac{1}{2m}(p_1{}^2 + p_2{}^2 + p_3{}^2). \tag{1.103}$$

---

[23] From Eq. (1.80) [or (1.85)], we have

$$\boldsymbol{p} = \frac{\partial L}{\partial \boldsymbol{v}} = m\boldsymbol{v} \quad \to \quad \boldsymbol{v} = \frac{\boldsymbol{p}}{m}, \quad \text{and} \quad L = \frac{1}{2}mv^2 = \frac{p^2}{2m}.$$

According to our recipe, we introduce the Hamilton's principal function W (with generating function of §§1.5.3 **b**-type). Since

$$p_i = \frac{\partial W}{\partial x_i}, \quad Q_i = \frac{\partial W}{\partial P_i}, \quad (i = 1, 2, 3),$$

the Hamilton–Jacobi equation (1.94) becomes

$$\frac{1}{2m}\left[\left(\frac{\partial W}{\partial x_1}\right)^2 + \left(\frac{\partial W}{\partial x_2}\right)^2 + \left(\frac{\partial W}{\partial x_3}\right)^2\right] + \frac{\partial W}{\partial t} = 0. \tag{1.104}$$

The latter equation is a type of §§1.4.4 **a**, *i.e.*, $f(p_i, p) = 0$, where $p = \partial W/\partial t$, so that the solution is given by putting $p_i = \alpha_i$ $(i = 1, 2, 3)$ and $p = -E$. Here, α_i and E are constants that satisfy

$$\frac{1}{2m}(\alpha_1{}^2 + \alpha_2{}^2 + \alpha_3{}^2) = E.$$

Using these constants, the solution is expressed as

$$W = \alpha_i x_i - Et + C$$
$$= \alpha_i x_i - \frac{1}{2m}(\alpha_1{}^2 + \alpha_2{}^2 + \alpha_3{}^2)\,t + C, \tag{1.105}$$

where C is an arbitrary constant.[24] Since our Hamilton–Jacobi equation includes only the partial derivatives, any constant is a trivial solution, which has no effect on the final results, and hence, we put $C = 0$ hereafter.

Now that the generating function (Hamilton's principal function) W is determined, we can calculate (following Eqs. (1.97), (1.98) and (1.99))

$$Q_i = \frac{\partial W}{\partial \alpha_i} = x_i - \frac{\alpha_i t}{m} = \beta_i \text{ (constant)},$$
$$\text{from which } x_i = \frac{\alpha_i t}{m} + \beta_i \ \big(= x_i(\boldsymbol{\alpha},\ \boldsymbol{\beta}, t)\big), \tag{1.106}$$
$$p_i = \frac{\partial W}{\partial x_i} = \alpha_i, \quad (i = 1, 2, 3).$$

If we rewrite $\alpha_i/m = v_i$, the above results are expressed as

$$x_i = v_i t + \beta_i, \quad p_i = m v_i, \tag{1.107}$$

By the Legendre transformation, we have

$$H = \boldsymbol{p} \cdot \boldsymbol{v} - L = \frac{p^2}{2m}.$$

---

[24]The same result may be obtained by assuming the function W of the form $W = U(x_1) + U(x_2) + U(x_3) + T(t)$, substituting it for Eq. (1.104), and the separation-of-variable-type calculation is made.

which describe a linear uniform motion of the particle. In the inertial frame of reference moving with a constant velocity v_i, we have

$$Q_i = x_i - v_i t = \beta_i \,(\text{constant}), \quad (i = 1, 2, 3), \tag{1.108}$$

which shows that the particle is observed at rest in this frame of reference.

$\triangleleft$

Chapter 2

Second-order Partial Differential Equations

In this chapter, we deal with the basics of the classification and the method of solutions on second-order linear partial differential equations (PDEs), which are important in science and engineering. In particular, the method of the separation of variables to reduce the PDEs to ODEs, the method of orthogonal function expansions, and the method of solving the initial-value and/or boundary-value problems by the use of Green's function, are focused as powerful tools to obtain the solution of PDEs. Another important method of solution by the use of the integral transforms will be developed in the next chapter.

2.1 Examples of Second-order PDEs

Many PDEs that appear in science and engineering are second-order and linear. A second-order differential equation implies a relation between any number of independent variables, a dependent variable depending on such independent variables, and the utmost second partial derivatives with respect to them. In a linear differential equation, the superposition of two or more solutions (if any) also gives the solution of the same differential equation. Keeping these fundamental properties in mind, we first look in detail the three types of examples, wave equation, diffusion equation, and the Poisson equation (including the Laplace equation as a special case). In particular, we pay attention to the derivation of the PDEs in this section, and the details on the method for solving them will be left for later sections.

2.1.1 *Wave equation*

Consider the vibration of a one-dimensional string. As shown in Fig. 2.1(a), we assume that both ends of the string are fixed to the walls with a distance

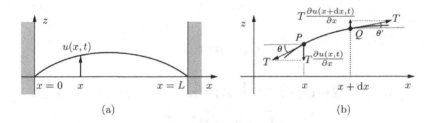

Fig. 2.1   (a) Vibration of a string with both ends fixed. (b) Tension acting on the line segment of the string.

L apart, at which the string is held by a tension T. Cartesian coordinates (x, z) with the x axis along the otherwise straight string are used. We consider a small line segment of the string $\mathrm{d}x$ at position x, and denote its displacement in the z direction at time t by $u(x, t)$. From the Newton's second law of motion, we have

$$(\rho\,\mathrm{d}x)\frac{\partial^2 u(x, t)}{\partial t^2} = T\left(\frac{\partial u(x + \mathrm{d}x, t)}{\partial x} - \frac{\partial u(x, t)}{\partial x}\right), \qquad (2.1)$$

where ρ is the line density (*i.e.*, mass of the string per unit length). Here, we take into account that the tension acts in the tangential direction at every position of the string, and only the z component of the equation of motion is shown (see Fig. 2.1(b)).[1] After expanding the r.h.s. of Eq. (2.1) in the Taylor's series up to the first order of $\mathrm{d}x$, we obtain

$$\frac{\partial^2 u(x, t)}{\partial t^2} = c^2 \frac{\partial^2 u(x, t)}{\partial x^2}, \qquad (2.2)$$

where we put $c = \sqrt{T/\rho}$. We see in the later subsection (§§2.2.2) that the quantity c has a meaning of the velocity. For this reason, Eq. (2.2) is called the **wave equation**, which is the fundamental equation for the wave phenomena in general, including light waves, acoustic waves, and electromagnetic waves. Note that this equation has symmetry with respect to the time reversal, so that it is invariant by the transformation $t \to -t$. The condition imposed at the end point of the string is called the **boundary condition**. In the present case, both ends are fixed, which is described as $u(0, t) = u(L, t) = 0$.

---

[1] At the left end of the line segment P, the z component of the tension is $-T\sin\theta \approx -T\tan\theta = -T\partial u(x, t)/\partial x$ for $\theta \ll 1$, whereas it is $T\sin\theta' \approx T\tan\theta' = T\partial u(x + \mathrm{d}x, t)/\partial x$ at the right end Q for $\theta' \ll 1$, which yields Eq. (2.1).

2.1.2 Diffusion equation

When a drop of ink falls into a cup of water, it spreads with time, and eventually fills the cup uniformly. We shall derive an equation that describes this process. For simplicity, we confine our attention to a one-dimensional case, and denote the density of the ink at position x at time t by $u(x,t)$. Consider the time variation of the amount of ink in a cylindrical region of a cross-sectional area S and length dx along the x axis. The increment of the ink in this region per unit time is given by $\partial/\partial t\,[u(x,t)Sdx]$. Meanwhile, we know, as a first approximation, that the flux[2] of a material $J(x,t)$ is proportional to the gradient of its density, with a direction from the higher density to the lower density sides, so that the flux of ink is described by

$$J(x,t) = -D\frac{\partial u(x,t)}{\partial x}, \tag{2.3}$$

where D is a positive constant. Then, the increment of the ink in this region is equal to the difference of the amount of ink flowing into this region through the area S at x and the one flowing out of this region through the area S at $x + dx$. Thus the relation

$$\frac{\partial}{\partial t}[u(x,t)(S\,dx)] = J(x,t)S - J(x+dx,t)S = -\frac{\partial J(x,t)}{\partial x}S\,dx \tag{2.4}$$

holds. Together with Eq. (2.3) and Eq. (2.4), we obtain

$$\frac{\partial u(x,t)}{\partial t} = D\frac{\partial^2 u(x,t)}{\partial x^2}, \tag{2.5}$$

where D is called the **diffusion constant**, and Eq. (2.5) is called the **diffusion equation**.

In contrast to the wave equation mentioned above, the diffusion equation is not invariant with the transformation $t \to -t$. This implies that if we reverse a movie film capturing the diffusion process of the ink, the concentration of ink to a point from a uniform distribution would be observed, which cannot occur in the real world, *i.e.*, the present equation shows an irreversible process. Note that the mathematical expression that the ink is concentrated at a point $x = 0$ at an initial instant $t = 0$ is given by $u(x,0) = \delta(x)$, where $\delta(x)$ is the delta function (as described later in §§2.4.1). The condition imposed at an initial instant is called the **initial condition**.

---

[2] Flux is the amount of material that passes through a unit area perpendicularly in a unit time.

2.1.3   Laplace–Poisson equation

Electrostatic potential problems are important in electromagnetism. For simplicity, we shall consider a two-dimensional case. We know that the electric field $\boldsymbol{E}$ is given by $\boldsymbol{E} = (E_x, E_y) = -\nabla u(x, y) = -(\partial u/\partial x, \partial u/\partial y)$, where $u(x, y)$ is called the scalar potential. From the Maxwell's equation, we have

$$\nabla \cdot \boldsymbol{E}(x, y) = \frac{\partial E_x(x, y)}{\partial x} + \frac{\partial E_y(x, y)}{\partial y} = \frac{\rho(x, y)}{\varepsilon_0}, \tag{2.6}$$

where $\rho(x, y)$ is the charge density, and ε_0 is the permittivity constant. In terms of u, the above equation becomes

$$\triangle u(x, y) \equiv \frac{\partial^2 u(x, y)}{\partial x^2} + \frac{\partial^2 u(x, y)}{\partial y^2} = f(x, y), \tag{2.7}$$

where we put $f(x, y) = -\rho(x, y)/\varepsilon_0$. Equation (2.7) is called the **Poisson equation**.

In particular, when $f = 0$, we have

$$\triangle u(x, y) \equiv \frac{\partial^2 u(x, y)}{\partial x^2} + \frac{\partial^2 u(x, y)}{\partial y^2} = 0, \tag{2.8}$$

which is called the **Laplace equation**, and its solution is called the **harmonic function**.

In general, second derivatives with respect to the spatial variable characterizes the curvature.[3] Therefore, the two-dimensional Laplace equation implies that the sum of the curvature in the x direction and that in the y direction is zero at every point on the integral surface. This can be seen, *e.g.*, in the shape of a soap film covering a closed wire frame. If the surface is convex in one direction, then it is concave in the perpendicular direction at every point on the soap surface (see Fig. 2.2). The pressure difference between both sides of the film is zero. However, if the soap film closes itself with a convex surface, similar to a soap bubble, the sum of the curvatures is not zero. This means that the pressure inside the soap bubble is higher than that outside. This situation is similar to that in electromagnetism. The electrostatic potential in the absence of a charge obeys the Laplace

---

[3]The curvature κ of a function $y = u(x)$ is given by

$$\kappa = \frac{u''(x)}{(1 + u'^2)^{3/2}},$$

the reciprocal of which is the radius of curvature R. For slowly varying functions $u'^2 \ll 1$, so that $\kappa \approx u''(x)$.

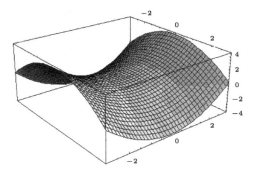

Fig. 2.2   Laplace equation and integral surface.

equation, and the equi-potential surface is a smooth surface with zero average curvature. However, the electrostatic potential enclosing a charge is governed by the Poisson equation, reflecting the presence of the electric field inside the surface.

2.2   Classification of Second-order PDEs

2.2.1   *Classification (hyperbolic, elliptic, parabolic)*

The general form of second-order linear PDEs for $u(x, y)$, where u is the variable depending on two independent variables x and y, is given by

$$A\frac{\partial^2 u}{\partial x^2} + 2B\frac{\partial^2 u}{\partial x \partial y} + C\frac{\partial^2 u}{\partial y^2} + D\frac{\partial u}{\partial x} + E\frac{\partial u}{\partial y} + Fu = G. \qquad (2.9)$$

Here, $A, B, \cdots$ and their partial derivatives are the continuous known functions of x and y. We assume that A, B, and C are not simultaneously zero ($A^2 + B^2 + C^2 \neq 0$). To simplify the form of this equation, we consider a coordinate transformation $(x, y) \rightarrow (\xi, \eta)$ *i.e.*, $\xi = \xi(x, y)$, $\eta = \eta(x, y)$. Then, Eq. (2.9) is generally rewritten as

$$A^*\frac{\partial^2 u}{\partial \xi^2} + 2B^*\frac{\partial^2 u}{\partial \xi \partial \eta} + C^*\frac{\partial^2 u}{\partial \eta^2} + D^*\frac{\partial u}{\partial \xi} + E^*\frac{\partial u}{\partial \eta} + F^*u = G^*, \qquad (2.10)$$

where

$$A^* = A\xi_x{}^2 + 2B\xi_x\xi_y + C\xi_y{}^2, \qquad (2.11)$$

$$B^* = A\xi_x\eta_x + B(\xi_x\eta_y + \xi_y\eta_x) + C\xi_y\eta_y, \qquad (2.12)$$

$$C^* = A\eta_x{}^2 + 2B\eta_x\eta_y + C\eta_y{}^2, \qquad (2.13)$$

$$D^* = A\xi_{xx} + 2B\xi_{xy} + C\xi_{yy} + D\xi_x + E\xi_y, \qquad (2.14)$$

$$E^* = A\eta_{xx} + 2B\eta_{xy} + C\eta_{yy} + D\eta_x + E\eta_y, \qquad (2.15)$$

$$F^* = F\big(x(\xi,\eta), y(\xi,\eta)\big), \qquad (2.16)$$

$$G^* = G\big(x(\xi,\eta), y(\xi,\eta)\big), \qquad (2.17)$$

$$B^{*2} - A^*C^* = (\xi_x\eta_y - \xi_y\eta_x)^2(B^2 - AC). \qquad (2.18)$$

We shall choose ξ and η, so that A^*, B^*, C^* become appropriate for characterizing the PDEs.[4] Note that the subscripts x, y used in this subsection, such as ξ_x, ξ_y, η_x, $\cdots$, refer to partial differentiations with respect to the relevant variables, and do not show the components of a vector quantity.

By rewriting the coefficients A^* and C^*

$$\frac{A^*}{\xi_y{}^2} = A\left(\frac{\xi_x}{\xi_y}\right)^2 + 2B\left(\frac{\xi_x}{\xi_y}\right) + C, \qquad (2.19)$$

$$\frac{C^*}{\eta_y{}^2} = A\left(\frac{\eta_x}{\eta_y}\right)^2 + 2B\left(\frac{\eta_x}{\eta_y}\right) + C, \qquad (2.20)$$

and comparing with the solution $z(A, B, C)$ of the quadratic equation $Az^2 + 2Bz + C = 0$, it is shown that Eq. (2.10) is classified into the following three types:

a. Case $B^2 - AC > 0$

In this case, $Az^2 + 2Bz + C = 0$ has two real solutions. If we choose ξ_x/ξ_y and η_x/η_y as the real solutions of this quadratic equation, we can put $A^* = 0$ and $C^* = 0$. Then, Eq. (2.10) becomes

$$2B^*\frac{\partial^2 u}{\partial\xi\partial\eta} + D^*\frac{\partial u}{\partial\xi} + E^*\frac{\partial u}{\partial\eta} + F^*u = G^*, \qquad (2.21)$$

or

$$\frac{\partial^2 u}{\partial\xi\partial\eta} = P\left(\xi, \eta, u, \frac{\partial u}{\partial\xi}, \frac{\partial u}{\partial\eta}\right). \qquad (2.22)$$

---

[4]In the transformation from the old set of variables (x, y) to the new set (ξ, η), the necessary and sufficient condition to assure the one-to-one correspondence between these two sets is the non-zero Jacobian $J \neq 0$, where $J \equiv \xi_x\eta_y - \xi_y\eta_x$. In the above transformation, $J \neq 0$ is assumed in the neighborhood of the point in question.

If we further transform the variables to $\xi = (1/2)(s+ct)$ and $\eta = (1/2)(s-ct)$, we obtain

$$\frac{\partial^2 u}{\partial s^2} - \frac{1}{c^2}\frac{\partial^2 u}{\partial t^2} = \tilde{P}\left(s, t, u, \frac{\partial u}{\partial s}, \frac{\partial u}{\partial t}\right). \tag{2.23}$$

These PDEs are called the standard form of the **hyperbolic PDEs**.

The PDE often used in science and engineering is

$$\frac{\partial^2 u}{\partial x^2} = \frac{1}{c^2}\frac{\partial^2 u}{\partial t^2}, \tag{2.24}$$

which is called a **wave equation** from the view point of physics (see §§2.1.1). In particular, the PDE that depends on one spatial variable x is a one-dimensional wave equation. When the phenomenon depends on two spatial variables x, y, as demonstrated in the vibration of the membrane, the governing equation becomes a two-dimensional wave equation:

$$\frac{\partial^2 u}{\partial x^2} + \frac{\partial^2 u}{\partial y^2} = \frac{1}{c^2}\frac{\partial^2 u}{\partial t^2}. \tag{2.25}$$

Similarly, for the problems dealing with a three-dimensional extension, such as the pressure waves, seismic waves, electromagnetic waves, etc. that are generated at a three-dimensional source point, the waves depend on the spatial variables x, y, z, so that the governing equation becomes

$$\frac{\partial^2 u}{\partial x^2} + \frac{\partial^2 u}{\partial y^2} + \frac{\partial^2 u}{\partial z^2} = \frac{1}{c^2}\frac{\partial^2 u}{\partial t^2}. \tag{2.26}$$

Example 2.1 (Derivation of Eq. (2.22)). With constants A, B, C given, how can the variables x, y be transformed to ξ, η to reduce the original PDE to Eq. (2.22)?

For the present purpose, we need to have $A^* = 0$ and $C^* = 0$. To realize the former, ξ should satisfy

$$A\xi_x + \left(B + \sqrt{B^2 - AC}\right)\xi_y = 0 \quad \text{or} \quad A\xi_x + \left(B - \sqrt{B^2 - AC}\right)\xi_y = 0,$$

from Eq. (2.11) [or, Eq. (2.19)]. The above PDEs are of the same form as that given in Chapter 1 (Eq. (1.19), §§1.2.2) with the choices of

$$a = A, \quad b = B \pm \sqrt{B^2 - AC}, \quad p = \frac{\partial \xi}{\partial x}, \quad q = \frac{\partial \xi}{\partial y}.$$

The solutions of the above equations are given by Eq. (1.21):

$$(B \pm \sqrt{B^2 - AC})x - Ay = \text{constant}.$$

The condition $C^* = 0$ for the variable η is similarly obtained. Taking into account that ξ and η refer to different coordinates, we may choose, *e.g.*,

$$\xi = (B + \sqrt{B^2 - AC})x - Ay, \quad \eta = (B - \sqrt{B^2 - AC})x - Ay. \tag{2.27}$$

◁

b. Case $B^2 - AC < 0$

In this case, the above-mentioned quadratic equation for z has two complex conjugate solutions. By choosing them for $\xi_x/\xi_y, \eta_x/\eta_y$, we can make $A^* = 0$ and $C^* = 0$. Then, we have from Eq. (2.10)

$$\frac{\partial^2 u}{\partial\xi\partial\eta} = Q\left(\xi, \eta, u, \frac{\partial u}{\partial\xi}, \frac{\partial u}{\partial\eta}\right), \tag{2.28}$$

following the same procedures as in **a**. However, taking account of the relation $\eta = \bar{\xi}$, we further transform $\xi = (1/2)(\sigma + i\tau)$, $\eta = (1/2)(\sigma - i\tau)$, which yields

$$\frac{\partial^2 u}{\partial\sigma^2} + \frac{\partial^2 u}{\partial\tau^2} = \tilde{Q}\left(\sigma, \tau, u, \frac{\partial u}{\partial\sigma}, \frac{\partial u}{\partial\tau}\right). \tag{2.29}$$

The latter type of equations are called the standard form of the **elliptic PDEs**.

One of the PDEs often used in science and engineering is

$$\frac{\partial^2 u}{\partial x^2} + \frac{\partial^2 u}{\partial y^2} = 0, \tag{2.30}$$

which is the **Laplace equation** shown in Eq. (2.8). To clarify the dependence on spatial variables x and y, it is sometimes called the two-dimensional Laplace equation, and its dependent variable u is simply called the "potential". The above PDE with non-zero r.h.s.:

$$\frac{\partial^2 u}{\partial x^2} + \frac{\partial^2 u}{\partial y^2} = g(x, y) \tag{2.31}$$

is called the **Poisson equation** (see §§2.1.3). In the three-dimensional problems depending on three spatial variables x, y, z, Eq. (2.30) and Eq. (2.31) are extended to $\triangle u = 0$ and $\triangle u = g$, respectively. Here, $\triangle$ defined by

$$\triangle = \frac{\partial^2}{\partial x^2} + \frac{\partial^2}{\partial y^2} + \frac{\partial^2}{\partial z^2}, \tag{2.32}$$

is called the Laplace operator, or the **Laplacian**.

c. Case $B^2 - AC = 0$

In this case, the above-mentioned quadratic equation for z has one double root, so that we can choose either $A^* = 0$ or $C^* = 0$. We shall put $C^* = 0$, for instance. Furthermore, if we put $\xi_x/\xi_y = \eta_x/\eta_y$ in Eq. (2.18), then we have $B^{*2} - A^*C^* = (\xi_x\eta_y - \xi_y\eta_x)^2(B^2 - AC) = 0$, which yields $B^* = 0$. Consequently, Eq. (2.10) becomes

$$A^*\frac{\partial^2 u}{\partial\xi^2} + D^*\frac{\partial u}{\partial\xi} + E^*\frac{\partial u}{\partial\eta} + F^*u = G^*, \tag{2.33}$$

or

$$\frac{\partial^2 u}{\partial \xi^2} = R\left(\xi, \eta, u, \frac{\partial u}{\partial \xi}, \frac{\partial u}{\partial \eta}\right). \tag{2.34}$$

The latter type of equations are called the standard form of the **parabolic PDEs**.

The PDE often used in science and engineering:

$$\frac{\partial u}{\partial t} = D\frac{\partial^2 u}{\partial x^2}, \tag{2.35}$$

is called the **diffusion equation** or the **heat conduction equation** (see §§2.1.2).

2.2.2 *Integral surfaces and initial-value problems*

a. Integral surfaces and initial-value problems

In the previous chapter (§§1.2.1), we have dealt with an integral surface of the first-order PDEs [Eq. (1.8)] for u that depends on two independent variables x, y, which can be extended to the function depending on many more variables as well as to the higher order PDEs. Namely, if the variable u depends on $x, y, \ldots, t$, and their relation is given by the PDEs, the solution $u = \psi(x, y, \ldots, t)$ shows a certain surface in the $xy \ldots tu$ space. The latter surface is also called the **integral surface**. Similarly to the one mentioned in the previous chapter (§1.3), the problem of finding the solution of higher-order PDEs that satisfies the conditions given at an initial instant is called the **initial-value problem**. Note that in higher-order PDEs, the initial values include not only the values of the function, but also their partial derivatives.

In the following, we confine our attention to a second-order PDE for u that depends on two independent variables x, y. Then, our *initial value problem* is to "find the solution of a given PDE:

$$Au_{xx} + 2Bu_{xy} + Cu_{yy} + Du_x + Eu_y + Fu = G, \tag{2.36}$$

that satisfies the condition

$$u = f(\sigma), \qquad \frac{\partial u}{\partial n} = g(\sigma) \tag{2.37}$$

on the curve l_0 in the xy plane."[5]

---

[5]Note that Eq. (2.36) is the same as Eq. (2.9) except for the difference in notation. Here, functions $A, B, \cdots$ (with $A^2 + B^2 + C^2 \neq 0$) and their partial derivatives are assumed to be continuous known functions of x and y.

As shown in Fig. 1.3, σ is the coordinate (parameter) along the curve l_0, and the point Q on this curve l_0 is given by $x = \xi(\sigma)$ and $y = \eta(\sigma)$, whereas the point P (which corresponds to the point Q on the integral surface) is given by $\big(\xi(\sigma), \eta(\sigma), u(\xi, \eta)\big)$. Here, $\partial/\partial n$ is the differentiation in the direction perpendicular to the curve l_0. In general, the condition in which the *values* of the dependent variables and their *normal gradient* are specified on the boundary curve in question (irrespective of the time variable or space variable) is called the **Cauchy condition**, and the problems of obtaining solutions that satisfies the Cauchy condition is called the **Cauchy's problem**.

In the Cauchy's problem of a first-order PDE, the condition for the value u is specified on the initial curve in the xy plane. However, in the Cauchy's problem of a second-order PDE, the conditions for u and p, q (and hence the information of the tangent plane) are specified on the initial curve in the xy plane.

b. Initial-value problems and power series expansion

To calculate u in the neighborhood of the curve l_0, we perform the following double Taylor series expansion:

$$u(x,y) = u(\xi, \eta) + [(x - \xi)u_x + (y - \eta)u_y] +$$
$$+ \frac{1}{2}[(x - \xi)^2 u_{xx} + 2(x - \xi)(y - \eta)u_{xy} + \cdots] + \cdots, \qquad (2.38)$$

where u and its partial derivatives of the r.h.s. are to be evaluated at a point $Q\big(\xi(\sigma), \eta(\sigma)\big)$ on the curve l_0. If the latter calculation is made, we can uniquely determine $u(x, y)$ within a radius of convergence around the point Q. Furthermore, if a similar calculation is made successively along the curve l_0, the value of u is uniquely determined in the region adjacent to the curve l_0.

For the above-mentioned purpose, we need to evaluate all quantities $u_x (= p), u_y (= q), u_{xx}, \cdots$ on the curve l_0 as functions of σ. Consider first p and q. Taking into account that the unit vector $\boldsymbol{t}_1$ tangent to the curve l_0, and the unit vector $\boldsymbol{n}_1$ normal to the latter, are given by (see Fig. 1.3)

$$\boldsymbol{t}_1 = (\dot{\xi}, \dot{\eta}), \qquad \boldsymbol{n}_1 = (\dot{\eta}, -\dot{\xi}), \qquad \text{where} \qquad \dot{} = \frac{\mathrm{d}}{\mathrm{d}\sigma},$$

we obtain

$$\frac{\partial u}{\partial \sigma} = \nabla u \cdot \boldsymbol{t}_1 = u_x \dot{\xi} + u_y \dot{\eta} = p\dot{\xi} + q\dot{\eta},$$

$$\frac{\partial u}{\partial n} = \nabla u \cdot \boldsymbol{n}_1 = u_x \dot{\eta} - u_y \dot{\xi} = p\dot{\eta} - q\dot{\xi},$$

where $\partial/\partial\sigma$ and $\partial/\partial n$ denote partial derivatives in the direction of $\boldsymbol{t}_1$ and $\boldsymbol{n}_1$, respectively. From these equations, we obtain

$$p = \dot{\xi}\frac{\partial u}{\partial\sigma} + \dot{\eta}\frac{\partial u}{\partial n} = \dot{\xi}\dot{f} + \dot{\eta}g, \qquad q = \dot{\eta}\frac{\partial u}{\partial\sigma} - \dot{\xi}\frac{\partial u}{\partial n} = \dot{\eta}\dot{f} - \dot{\xi}g. \tag{2.39}$$

The r.h.s. are known functions of σ, and hence are always calculable.

How do we consider the partial derivatives of p and q, *i.e.*, u_{xx}, u_{xy}, u_{yy}, $\cdots$? To do so, we calculate, *e.g.*,

$$\frac{\partial p}{\partial\sigma} = \frac{\partial p}{\partial x}\frac{\partial x}{\partial\sigma} + \frac{\partial p}{\partial y}\frac{\partial y}{\partial\sigma} = u_{xx}\,\dot{x} + u_{xy}\,\dot{y},$$

and evaluate it at the point $Q(\xi,\eta)$, which results in

$$\dot{\xi}\,u_{xx} + \dot{\eta}\,u_{xy} = \frac{\partial p}{\partial\sigma}. \tag{2.40}$$

Similarly, we have from $\partial q/\partial\sigma$,

$$\dot{\xi}\,u_{xy} + \dot{\eta}\,u_{yy} = \frac{\partial q}{\partial\sigma}. \tag{2.41}$$

The r.h.s. of Eqs. (2.40), (2.41), as well as $\dot{\xi},\dot{\eta}$, are given at the initial points, and are hence known. These equations and the PDE (2.36):

$$A(\sigma)\,u_{xx} + 2B(\sigma)\,u_{xy} + C(\sigma)\,u_{yy} = G(\sigma) - D(\sigma)\,u_x - E(\sigma)\,u_y - F(\sigma)\,u \tag{2.42}$$

constitute a system of linear equations for three variables u_{xx}, u_{xy}, u_{yy}, which is solved uniquely if the Jacobian

$$\Delta \equiv \begin{vmatrix} \dot{\xi} & \dot{\eta} & 0 \\ 0 & \dot{\xi} & \dot{\eta} \\ A & 2B & C \end{vmatrix} = A\,\dot{\eta}^2 - 2B\,\dot{\xi}\dot{\eta} + C\,\dot{\xi}^2 \tag{2.43}$$

is not zero. In other words, if $\Delta \neq 0$ holds at any points on the curve l_0, then the second-order partial derivatives are determined uniquely along this curve. Similarly, higher-order partial derivatives are successively determined. In this way, we can extend the region of validity starting from the points on the curve l_0, and hence the initial-value problem (Cauchy's problem) is solved.

c. Initial-value problems and characteristic curves

When the Jacobian given by Eq. (2.43) is zero (*i.e.*, $\Delta = 0$), u_{xx}, u_{xy}, u_{yy} are linearly dependent, so that the solution is not obtained by the method shown in the previous subsection §§2.2.2b. In some cases, however, solutions can be obtained by the method of **characteristic curve** developed in the previous chapter (§§1.2.2).

We first remark that the relation $\Delta = 0$ is the alternative form of the characteristic curve (characteristic differential equation) in the xy plane:

$$A\,(\mathrm{d}y)^2 - 2B\,\mathrm{d}x\,\mathrm{d}y + C\,(\mathrm{d}x)^2 = 0, \tag{2.44}$$

which can be shown as follows. In accordance with the classification of §§2.2.1, we examine three cases depending on the values of A, B, C.

(a) Case $B^2 - AC > 0$

As stated regarding the transformation of independent variables from x, y to ξ, η (§§2.2.1), if we choose ξ_x/ξ_y for the solution of the quadratic equation $Az^2 + 2Bz + C = 0$, then A^* becomes zero, which accordingly requires $A\xi_x^{\,2} + 2B\xi_x\xi_y + C\xi_y^{\,2} = 0$, i.e.,

$$A\,\xi_x + (B + \sqrt{B^2 - AC}\,)\xi_y = 0, \quad \text{or} \quad A\,\xi_x + (B - \sqrt{B^2 - AC}\,)\xi_y = 0. \tag{2.45}$$

Each of these equations is a first-order PDE for $\xi(x, y)$, which is reminiscent of the characteristic curve. Indeed, on the curve "$\xi(x, y) = \text{constant}$" that satisfies either one of the Eqs. (2.45), we have

$$\xi_x\,\mathrm{d}x + \xi_y\,\mathrm{d}y = 0, \tag{2.46}$$

so that the relations derived from Eqs. (2.45) and (2.46):

$$\frac{\mathrm{d}x}{A} = \frac{\mathrm{d}y}{B \pm \sqrt{B^2 - AC}},$$

or the equivalent expressions:

$$A\,\mathrm{d}y - (B + \sqrt{B^2 - AC}\,)\,\mathrm{d}x = 0, \quad A\,\mathrm{d}y - (B - \sqrt{B^2 - AC}\,)\,\mathrm{d}x = 0, \tag{2.47}$$

give the characteristic directions. By multiplying these two equations, we obtain Eq. (2.44).

The integration of Eq. (2.47) yields the characteristic curves

$$\varphi_1(x, y) \equiv A\,y - (B + \sqrt{B^2 - AC}\,)\,x = \text{constant},$$
$$\varphi_2(x, y) \equiv A\,y - (B - \sqrt{B^2 - AC}\,)\,x = \text{constant}, \tag{2.48}$$

along which the initial data specified at the point $Q(\xi(\sigma), \eta(\sigma))$ on the initial curve l_0 are transmitted to any points on these characteristic curves. This implies that $\dot{\xi} \equiv \mathrm{d}\xi/\mathrm{d}\sigma$ and $\dot{\eta} \equiv \mathrm{d}\eta/\mathrm{d}\sigma$ are the same as $\mathrm{d}x/\mathrm{d}\sigma$ and $\mathrm{d}y/\mathrm{d}\sigma$, respectively, along the characteristic curves. Accordingly, Eq. (2.44) is rewritten as

$$A\left(\frac{\mathrm{d}y}{\mathrm{d}\sigma}\right)^2 - 2B\frac{\mathrm{d}x}{\mathrm{d}\sigma}\frac{\mathrm{d}y}{\mathrm{d}\sigma} + C\left(\frac{\mathrm{d}x}{\mathrm{d}\sigma}\right)^2 = 0 \quad \text{or} \quad A\,\dot{\eta}^2 - 2B\,\dot{\xi}\dot{\eta} + C\,\dot{\xi}^2 = 0, \tag{2.49}$$

which reproduces the r.h.s. of Eq. (2.43). Similar results are obtained for the curve "$\eta =$ constant" derived from the choice of $C^* = 0$, and hence, we come to the conclusion that there are two independent characteristic curves for the case $B^2 - AC > 0$.

If the curve l_0, on which the initial conditions are imposed, intersects the respective curves (2.48) only once, the solution $u(x, y)$ that satisfies the Cauchy condition is determined uniquely, because the value of $u(x, y)$ on the curve (2.48) is the same as the one at the intersecting point $Q(\xi, \eta)$ on the curve l_0. By repeating the procedures starting from an arbitrary point on the curve l_0, we can extend the solution that satisfies the initial condition on the curve l_0 to the entire region with $B^2 - AC > 0$. In this way, we can solve the initial value problems even if $\Delta = 0$.

(b) Case $B^2 - AC = 0$

In this case, the two equations of (2.47) coincide, so that only one characteristic curve is available, which is given by solving

$$A \, dy - B \, dx = 0. \tag{2.50}$$

We can choose this solution as a family of curves "$\xi(x, y) =$ constant." If the initial curve l_0 intersects the characteristic curve, we can solve the initial value problem similarly to the one mentioned in the previous case (a).

(c) Case $B^2 - AC < 0$

In this case, no real characteristic curve exists.   ◁

Example 2.2 (wave propagation). We shall consider the characteristic curves of Eq. (2.2), which are frequently met in science and engineering.

If we rewrite Eq. (2.2) as

$$c^2 \frac{\partial^2 u(x, t)}{\partial x^2} - \frac{\partial^2 u(x, t)}{\partial t^2} = 0,$$

and compare it with a general expression of the PDE (2.9), the variable y corresponds to t, and $A = c^2$, $B = 0$, $C = -1$, so that the present example belongs to the case $B^2 - AC > 0$. From Eq. (2.48), the characteristic curves are found to be

$$x - ct = C_1, \quad x + ct = C_2, \quad (C_1, \ C_2 \text{ constant}).$$

We will see in later subsections (§§2.5.2 and §§3.3.2) that the general solution of Eq. (2.2) is given by

$$u(x, t) = F(x - ct) + G(x + ct), \tag{2.51}$$

Partial Differential Equations

where F and G are arbitrary functions, which is called the **d'Alembert's solution** derived in 1747.

From a view point of physics, both of these are the progressive waves. For instance, consider the first term of r.h.s. $u(x, t) = F(x - ct)$. It shows that $u = F(x)$ at $t = 0$, which implies that the displacement at $t = 0$, $x = x_0$ is $u(x_0, 0) = F(x_0)$. After time Δt, the displacement at $x = x_0 + c\Delta t$ is $u(x_0 + c\Delta t, \Delta t) = F\big((x_0 + c\Delta t) - c(\Delta t)\big) = F(x_0)$, which means that the position with the same displacement moves toward the positive x direction by the amount $c\Delta t$ within the time interval Δt (see Fig. 2.3(a)). The same displacement is observed at any point x, not solely restricted to the position x_0 in question. Consequently, $F(x - ct)$ shows the waves propagating toward the positive x direction at a speed c. Similarly, we can show that $u(x, t) = G(x + ct)$ describes the waves propagating toward the negative x direction at a speed c.

Let us see this in Fig. 2.3(b). For example, at a point x, t on the curve (or the straight line in this case) $x - ct = a$ (constant), u takes a constant value $F(a)$, so that the displacement u propagates along this curve in the xt plane. Similarly, the point x, t that satisfies $x + ct = b$ (constant) propagates the displacement $u = G(b)$ along the latter curve. These curves are the characteristic curves, along which the displacement (or some type of information) at the marked point "propagates." In contrast to the first-order PDE dealt with in §§1.2.2, two characteristic curves appear in the present second-order PDE case, so that the event occurring at a certain point $(x_0, 0)$ in the xt plane propagates along the lines with gradient $\pm 1/c$, such as $A \to B$ and $A \to C$. At the same time, the propagation occurs *e.g.*, along $A \to D$, and then along $D \to E$, where the point D serves as a new wave source point. As a result, the event occurring at $x = x_0$ at a time

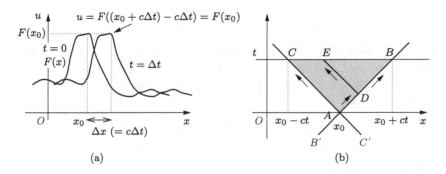

(a) (b)

Fig. 2.3   (a) Propagation of the wave $u = F(x - ct)$ and (b) characteristic curves.

$t = 0$ propagates to all the points within the wedge-shaped region ABC at time t (Fig. 2.3(b)). In this sense, the region ABC shown in this figure is called the "**region of influence**" of the event that occurred at the point $A(x_0, 0)$, whereas all the points in the region $AB'C'$ ($t < 0$ side) transmit the information to the point A and determine the event at $A(x_0, 0)$, so that the latter region is called the "**region of determination (or region of dependence)**". These relations are called "causality".   ◁

2.3 Separation of Variables and Eigenvalue Problems

Many problems in science and engineering are reduced to finding the solutions of the following PDEs.
 • **Laplace equation:**

$$\triangle \Psi = 0, \tag{2.52}$$

 • **diffusion equation** or **heat conduction equation:**

$$\triangle \Psi = \frac{1}{D}\frac{\partial \Psi}{\partial t}, \tag{2.53}$$

 • **wave equation:**

$$\triangle \Psi = \frac{1}{c^2}\frac{\partial^2 \Psi}{\partial t^2}, \tag{2.54}$$

where D and c are constants, and

$$\triangle \equiv \frac{\partial^2}{\partial x^2} + \frac{\partial^2}{\partial y^2} + \frac{\partial^2}{\partial z^2} \tag{2.55}$$

is the Laplacian. If the quantity Ψ has dependence on time t like $\exp(-\mu t)$ or $\cos(\omega t)$, $\sin(\omega t)$, Eq. (2.53) and Eq. (2.54) are reduced to the following type (where μ, ω and κ are assumed to be constants):

$$\triangle \Psi + \kappa \Psi = 0, \tag{2.56}$$

which is called the **Helmholtz equation**. The latter includes Eqs. (2.52)–(2.54) by the choice of κ, so that we shall focus to Eq. (2.56) in this section as a typical second-order linear PDE.

One of the methods for obtaining an accurate solution is to find an appropriate coordinate system

$$u_1 = f_1(x, y, z), \qquad u_2 = f_2(x, y, z), \qquad u_3 = f_3(x, y, z), \tag{2.57}$$

by which the solution of the given PDE is sought in the form of the product of functions depending separately on the respective variables[6]

$$\Psi = \Psi_1(u_1)\,\Psi_2(u_2)\,\Psi_3(u_3). \tag{2.58}$$

Furthermore, if $\triangle$ is separable (the meaning of which will be made clear later), then Eq. (2.56) is reduced to a set of three ODEs, each of which depends only on u_i, and hence is easily solvable. The latter is called the method of **separation of variables**. In the following, we examine the possible coordinate systems that satisfy these requirements.

2.3.1 *Orthogonal curvilinear coordinates and separation of variables*

a. Curvilinear coordinates

Consider the relation between the given coordinates u_1, u_2, u_3 and the rectangular coordinates (or Cartesian coordinates) x, y, z:

$$x = x(u_1, u_2, u_3), \qquad y = y(u_1, u_2, u_3), \qquad z = z(u_1, u_2, u_3). \tag{2.59}$$

The infinitesimal vector components are

$$\mathrm{d}x = \frac{\partial x}{\partial u_1}\mathrm{d}u_1 + \frac{\partial x}{\partial u_2}\mathrm{d}u_2 + \frac{\partial x}{\partial u_3}\mathrm{d}u_3,$$

$$\mathrm{d}y = \frac{\partial y}{\partial u_1}\mathrm{d}u_1 + \frac{\partial y}{\partial u_2}\mathrm{d}u_2 + \frac{\partial y}{\partial u_3}\mathrm{d}u_3, \tag{2.60}$$

$$\mathrm{d}z = \frac{\partial z}{\partial u_1}\mathrm{d}u_1 + \frac{\partial z}{\partial u_2}\mathrm{d}u_2 + \frac{\partial z}{\partial u_3}\mathrm{d}u_3,$$

so that the line element $\mathrm{d}s = \sqrt{\mathrm{d}x^2 + \mathrm{d}y^2 + \mathrm{d}z^2}$ is given by

$$\mathrm{d}s^2 = g_{11}\,\mathrm{d}u_1{}^2 + g_{22}\,\mathrm{d}u_2{}^2 + g_{33}\,\mathrm{d}u_3{}^2$$
$$+ g_{12}\,\mathrm{d}u_1\,\mathrm{d}u_2 + g_{23}\,\mathrm{d}u_2\,\mathrm{d}u_3 + g_{31}\,\mathrm{d}u_3\,\mathrm{d}u_1, \tag{2.61}$$

where

$$g_{ij} = \varepsilon\left(\frac{\partial x}{\partial u_i}\frac{\partial x}{\partial u_j} + \frac{\partial y}{\partial u_i}\frac{\partial y}{\partial u_j} + \frac{\partial z}{\partial u_i}\frac{\partial z}{\partial u_j}\right), \qquad (i, j = 1, 2, 3). \tag{2.62}$$

Here, $\varepsilon = 1$ for $i = j$, and $\varepsilon = 2$ for $i \neq j$. The coefficient g_{ij} is a second rank tensor called a **metric**.

When $g_{ij} = 0$ for $i \neq j$ (see §§2.3.1**c**), $\mathrm{d}s^2$ becomes

$$\mathrm{d}s^2 = g_{11}\,\mathrm{d}u_1{}^2 + g_{22}\,\mathrm{d}u_2{}^2 + g_{33}\,\mathrm{d}u_3{}^2. \tag{2.63}$$

---

[6]Note that the suffixes in the present subsection refer to the components of the vector (and not to specify the variables to perform differentiations), unless otherwise stated.

In this case, u_i's are solved as

$$u_1 = u_1(x, y, z), \qquad u_2 = u_2(x, y, z), \qquad u_3 = u_3(x, y, z), \qquad (2.64)$$

and the surfaces of constant u_i $(i = 1, 2, 3)$ are mutually perpendicular to each other. Namely, the coordinates u_1, u_2, u_3 constitute the **orthogonal curvilinear coordinates**. The boundary shape is specified by one of the coordinates, which is by far the most important to consider the boundary-value problems, separation of variables, and eigenfunction expansions, *etc.* as shown later.

In the following, we confine our attention to the orthogonal curvilinear coordinates (and hence the case with $g_{ij} = 0$, $i \neq j$), and denote

$$g_{11} = h_1{}^2, \qquad g_{22} = h_2{}^2, \qquad g_{33} = h_3{}^2, \qquad (2.65)$$

for brevity. The coefficients h_1, h_2, h_3 are also called the scale factors. Then, the line elements in the respective coordinates are given by

$$ds_1 = h_1 \, du_1, \qquad ds_2 = h_2 \, du_2, \qquad ds_3 = h_3 \, du_3. \qquad (2.66)$$

In this case, operators such as the gradient, divergence and rotation (or curl), which are frequently used in a vector analysis, are expressed as

$$\operatorname{grad} \Psi = \left(\frac{1}{h_1} \frac{\partial \Psi}{\partial u_1}, \ \frac{1}{h_2} \frac{\partial \Psi}{\partial u_2}, \ \frac{1}{h_3} \frac{\partial \Psi}{\partial u_3} \right), \qquad (2.67)$$

$$\operatorname{div} \boldsymbol{v} = \frac{1}{h_1 h_2 h_3} \left[\frac{\partial}{\partial u_1} (h_2 h_3 v_1) + \frac{\partial}{\partial u_2} (h_3 h_1 v_2) + \frac{\partial}{\partial u_3} (h_1 h_2 v_3) \right], \quad (2.68)$$

$$(\operatorname{rot} \boldsymbol{v})_1 = \frac{1}{h_2 h_3} \left[\frac{\partial}{\partial u_2} (h_3 v_3) - \frac{\partial}{\partial u_3} (h_2 v_2) \right], \qquad (2.69)$$

where the second and the third components of rot $\boldsymbol{v}$ are given by the cyclic permutations of the suffixes $i = 1, 2, 3$ in Eq. (2.69). Note that we denote $\boldsymbol{v} = (v_{u_1}, v_{u_2}, v_{u_3}) = (v_1, v_2, v_3)$. Furthermore, the Laplacian is given by $\triangle = \operatorname{div} \operatorname{grad}$, so that we have

$$\triangle \Psi = \frac{1}{h_1 h_2 h_3} \left[\frac{\partial}{\partial u_1} \left(\frac{h_2 h_3}{h_1} \frac{\partial \Psi}{\partial u_1} \right) + \frac{\partial}{\partial u_2} \left(\frac{h_3 h_1}{h_2} \frac{\partial \Psi}{\partial u_2} \right) + \frac{\partial}{\partial u_3} \left(\frac{h_1 h_2}{h_3} \frac{\partial \Psi}{\partial u_3} \right) \right].$$
$$(2.70)$$
$$\triangleleft$$

b. Rectangular coordinates (Cartesian coordinates) $(\boldsymbol{x}, \boldsymbol{y}, \boldsymbol{z})$
    The choice of

$$u_1 = x, \qquad u_2 = y, \qquad u_3 = z \qquad (2.71)$$

yields $h_1 = h_2 = h_3 = 1$, so that we have (from Eq. (2.70))

$$\triangle \Psi = \frac{\partial^2 \Psi}{\partial x^2} + \frac{\partial^2 \Psi}{\partial y^2} + \frac{\partial^2 \Psi}{\partial z^2}. \tag{2.72}$$

The solution of the Helmholtz equation $(\triangle + \kappa)\Psi = 0$ is obtained in the separable form $\Psi = X(x)Y(y)Z(z)$, where

$$\frac{d^2 X}{dx^2} + \lambda X = 0, \quad \frac{d^2 Y}{dy^2} + \mu Y = 0, \quad \frac{d^2 Z}{dz^2} + \nu Z = 0, \tag{2.73}$$

and $\kappa = \lambda + \mu + \nu$. The Schrödinger equation for the wave function Ψ with the total energy E and the potential energy $U(x, y, z)$ (see *e.g.*, [Schiff (1949)]):

$$\triangle \Psi + (E - U)\Psi = 0 \tag{2.74}$$

also has a separable solution in the rectangular coordinates if the potential is given in the form $U = U_x(x) + U_y(y) + U_z(z)$.

c. Separable condition

The condition (2.63) requires

$$g_{ij} = 2 \left(\frac{\partial x}{\partial u_i} \frac{\partial x}{\partial u_j} + \frac{\partial y}{\partial u_i} \frac{\partial y}{\partial u_j} + \frac{\partial z}{\partial u_i} \frac{\partial z}{\partial u_j} \right) = 0 \tag{2.75}$$

for $i \neq j$ $(i, j = 1, 2, 3)$. To fulfill the condition (2.75), we examine the geometric relation between the coordinate planes in the xyz space and those in the $u_1 u_2 u_3$ space.

To do so, we first consider the surface in the xyz space in general:

$$F(\lambda, x, y, z) \equiv \frac{x^2}{\lambda - a_1} + \frac{y^2}{\lambda - a_2} + \frac{z^2}{\lambda - a_3} - 1, \tag{2.76}$$

where a_1, a_2, a_3 are constants with $0 < a_3 < a_2 < a_1$. Then, the points (x, y, z) that fulfill the equation $F = 0$ describe a family of

(i) confocal ellipsoids   for $a_1 < \lambda < \infty$,

(ii) confocal hyperboloids of one sheet for $a_2 < \lambda < a_1$,

(iii) confocal hyperboloids of two sheets for $a_3 < \lambda < a_2$.

For a fixed x, y, z, the equation $F = 0$ provides a cubic equation of λ, which has three real solutions u_1, u_2, u_3 $(u_3 < u_2 < u_1)$. The ordering of these solutions and the constants are

$$0 < a_3 < u_3 < a_2 < u_2 < a_1 < u_1 < \infty, \tag{2.77}$$

which is trivial by looking at $F(\infty) = -1$, and subsequently $F(a_1 \pm 0) = \pm\infty$, $F(a_2 \pm 0) = \pm\infty$, and $F(a_3 \pm 0) = \pm\infty$. Conversely,

$$F(u_i, x, y, z) = 0, \qquad \text{or} \qquad u_i = u_i(x, y, z), \qquad (i = 1, 2, 3) \qquad (2.78)$$

describe three surfaces, which give one solution point (x^2, y^2, z^2). The latter can be determined as follows: Taking into account that $F = 0$ is a cubic equation of λ, with the solution given by u_1, u_2, u_3, Eq. (2.76) should be expressed as

$$\frac{x^2}{\lambda - a_1} + \frac{y^2}{\lambda - a_2} + \frac{z^2}{\lambda - a_3} - 1 = -\frac{(\lambda - u_1)(\lambda - u_2)(\lambda - u_3)}{f(\lambda)}, \qquad (2.79)$$

where $f(\lambda) \equiv (\lambda - a_1)(\lambda - a_2)(\lambda - a_3)$. The multiplication of Eq. (2.79) with $f(\lambda)$ and the substitution of $\lambda = a_1, a_2, a_3$ yield

$$\begin{aligned} x^2 &= \frac{(u_1 - a_1)(u_2 - a_1)(u_3 - a_1)}{(a_1 - a_2)(a_1 - a_3)}, \\ y^2 &= \frac{(u_1 - a_2)(u_2 - a_2)(u_3 - a_2)}{(a_2 - a_3)(a_2 - a_1)}, \\ z^2 &= \frac{(u_1 - a_3)(u_2 - a_3)(u_3 - a_3)}{(a_3 - a_1)(a_3 - a_2)}. \end{aligned} \qquad (2.80)$$

Using these results, we can calculate the line element. To do so, *e.g.*, we may take the logarithm of the first equation of Eq. (2.80), and differentiate it with respect to x, which yields

$$\mathrm{d}x = \frac{x}{2}\left(\frac{\mathrm{d}u_1}{u_1 - a_1} + \frac{\mathrm{d}u_2}{u_2 - a_1} + \frac{\mathrm{d}u_3}{u_3 - a_1}\right). \qquad (2.81)$$

Similarly, $\mathrm{d}y$ and $\mathrm{d}z$ are obtained, and $\mathrm{d}s^2 = \mathrm{d}x^2 + \mathrm{d}y^2 + \mathrm{d}z^2$ is calculated. The latter is given in the form

$$4\mathrm{d}s^2 = \left[\frac{x^2}{(u_1 - a_1)^2} + \frac{y^2}{(u_1 - a_2)^2} + \frac{z^2}{(u_1 - a_3)^2}\right]\mathrm{d}u_1{}^2 + 2\left[\frac{x^2}{(u_1 - a_1)(u_2 - a_1)}\right.$$
$$\left. + \frac{y^2}{(u_1 - a_2)(u_2 - a_2)} + \frac{z^2}{(u_1 - a_3)(u_2 - a_3)}\right]\mathrm{d}u_1\mathrm{d}u_2 + \cdots . \qquad (2.82)$$

Note that the coefficient of $\mathrm{d}u_1{}^2$ of the first term of the r.h.s. of Eq. (2.82) is the same as the one, in which the l.h.s. of Eq. (2.79) is differentiated with respect to λ, and put $\lambda = u_1$ (and further multiplication of -1). The same calculation, made for the r.h.s. of Eq. (2.79), yields

$$\frac{(u_1 - u_2)(u_1 - u_3)}{f(u_1)}. \qquad (2.83)$$

Meanwhile, the coefficient of $du_1 du_2$ in the second term of Eq. (2.82) is equal to $F(u_1, x, y, z) - F(u_2, x, y, z)$ divided by $u_2 - u_1$, which is zero from Eq. (2.78). The above-mentioned results also apply to the coefficient of $du_2{}^2$, $du_3{}^2$, $du_2 du_3$, and $du_3 du_1$, so that the line element is given by

$$ds^2 = \frac{1}{4} \sum_{(1,2,3)} \frac{(u_1 - u_2)(u_1 - u_3)}{f(u_1)} \, du_1{}^2, \qquad (2.84)$$

where the summation $(1, 2, 3)$ implies that the cyclic permutations are taken. Comparing the above expression with Eqs. (2.63) and (2.65), we obtain the relation

$$h_1{}^2 = \frac{(u_1 - u_2)(u_1 - u_3)}{4f(u_1)}. \qquad (2.85)$$

Similarly, h_2 and h_3 are obtained by cyclic permutations of the suffixes 1, 2, 3 in h_1^2. In the present coordinate system, the Helmholtz equation $(\triangle + \kappa)\Psi = 0$ is, with the use of Eq. (2.70), expressed in the form

$$\sum_{(1,2,3)} \frac{\sqrt{f(u_1)}}{(u_1 - u_2)(u_1 - u_3)} \frac{\partial}{\partial u_1} \left(\sqrt{f(u_1)} \frac{\partial \Psi}{\partial u_1} \right) + \frac{\kappa}{4} \Psi = 0. \qquad (2.86)$$

If we substitute $\Psi = \Psi_1(u_1)\Psi_2(u_2)\Psi_3(u_3)$ for Eq. (2.86), and divide by Ψ, we obtain

$$\sum_{(1,2,3)} \frac{1}{(u_1 - u_2)(u_1 - u_3)} \left[\frac{\sqrt{f(u_1)}}{\Psi_1} \frac{d}{du_1} \left(\sqrt{f(u_1)} \frac{d\Psi_1}{du_1} \right) \right] + \frac{\kappa}{4} = 0. \quad (2.87)$$

The function inside the bracket [ ] is a function of u_1, which we shall define as $G_1(u_1)$. Similarly, $G_2(u_2)$ and $G_3(u_3)$ are defined by the functions with suffixes 2 and 3, respectively. Then, Eq. (2.87) becomes

$$G_1(u_1)(u_2 - u_3) + G_2(u_2)(u_3 - u_1) + G_3(u_3)(u_1 - u_2)$$
$$= \frac{\kappa}{4}(u_1 - u_2)(u_2 - u_3)(u_3 - u_1). \qquad (2.88)$$

By differentiation of Eq. (2.88) twice with respect to u_i, we have

$$G_i''(u_i) = -\frac{\kappa}{2} \qquad (i = 1, 2, 3), \qquad (2.89)$$

which is integrated to obtain

$$G_i(u_i) = -\frac{1}{4}(\kappa u_i{}^2 + \alpha_i u_i + \beta_i), \qquad (i = 1, 2, 3), \qquad (2.90)$$

where α_i and β_i are arbitrary constants. From the symmetry of Eq. (2.88), we have $\alpha_1 = \alpha_2 = \alpha_3$, and $\beta_1 = \beta_2 = \beta_3$. If we further rewrite $\alpha_i = \alpha$,

$\beta_i = \beta$, $u_i = \zeta$, $\Psi_i = w$, then $\Psi_i(u_i)$ is separated and is expressed in the following form:

$$\sqrt{f(\zeta)}\frac{\mathrm{d}}{\mathrm{d}\zeta}\left(\sqrt{f(\zeta)}\frac{\mathrm{d}w}{\mathrm{d}\zeta}\right) + \frac{1}{4}(\kappa\zeta^2 + \alpha\zeta + \beta)w = 0. \qquad (2.91)$$

By substitution of $f(\zeta) = (\zeta - a_1)(\zeta - a_2)(\zeta - a_3)$ into the above expression, we finally obtain the ODE, which $\Psi_i(u_i)$ ($i = 1, 2, 3$) should separately satisfy:

$$\frac{\mathrm{d}^2 w}{\mathrm{d}\zeta^2} + \frac{1}{2}\left(\frac{1}{\zeta - a_1} + \frac{1}{\zeta - a_2} + \frac{1}{\zeta - a_3}\right)\frac{\mathrm{d}w}{\mathrm{d}\zeta} + \frac{\kappa\zeta^2 + \alpha\zeta + \beta}{4(\zeta - a_1)(\zeta - a_2)(\zeta - a_3)}w = 0. \qquad (2.92)$$

We show some specific examples of the above-mentioned general theory.

d. Prolate spheroidal coordinates (ξ, η, φ)

In §§2.3.1c, we have examined the general tri-axial spheroid with radii $\sqrt{u_1 - a_1} < \sqrt{u_1 - a_2} < \sqrt{u_1 - a_3}$. As a special case, we consider the limiting case in which the semi-axes $\sqrt{u_1 - a_2}$ and $\sqrt{u_1 - a_1}$ are of equal length. To do so, we put $a_1 - a_2 = \varepsilon$, $a_1 - a_3 = l^2$, $u_1 = a_3 + l^2 u^2$, $u_2 = a_2 + \varepsilon \sin^2 \varphi$, $u_3 = a_3 + l^2 v^2$ ($u > 1$, $|v| < 1$) in Eq. (2.80), and take the limit $\varepsilon \to 0$, which yields

$$\begin{aligned} x &= l\sqrt{(u^2 - 1)(1 - v^2)}\,\cos\varphi, \\ y &= l\sqrt{(u^2 - 1)(1 - v^2)}\,\sin\varphi, \qquad (2.93) \\ z &= luv, \end{aligned}$$

$$(u \geq 1, \ |v| \leq 1, \ 0 \leq \varphi < 2\pi).$$

This coordinate system defines the prolate spheroidal coordinates with the z axis as a symmetric axis. The surfaces of constant u give a family of confocal ellipsoids of revolution, those of constant v give a family of confocal hyperboloids of revolution of two sheets, and those of constant φ give coaxial planes. Figure 2.4(a) shows the prolate spheroidal coordinate system, where the z axis is chosen on the q_2 axis, φ as the angle of rotation around the z axis, and the x axis on the q_1 axis in the meridian plane $\varphi = $ constant ($= 0$).

In the present limiting procedures, we have

$$\sqrt{f(u_1)} \to l^3 u(u^2 - 1), \quad \sqrt{f(u_2)} \to il\varepsilon\cos\varphi\sin\varphi, \quad \sqrt{f(u_3)} \to l^3 v(1 - v^2),$$

$$\mathrm{d}u_1 = 2l^2 u\,\mathrm{d}u, \quad \mathrm{d}u_2 = 2\varepsilon\cos\varphi\sin\varphi\,\mathrm{d}\varphi, \quad \mathrm{d}u_3 = 2l^2 v\,\mathrm{d}v,$$

so that the Helmholtz equation separated into respective coordinates of the

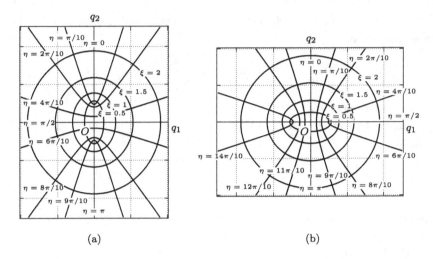

Fig. 2.4   (a) Prolate spheroidal coordinates, and (b) oblate spheroidal coordinates.

form $\Psi = U(u)V(v)\Phi(\varphi)$, given by Eq. (2.91), becomes

$$\frac{\mathrm{d}}{\mathrm{d}u}\left((u^2-1)\frac{\mathrm{d}U}{\mathrm{d}u}\right) + \left(\kappa l^2 u^2 - \Lambda - \frac{m^2}{u^2-1}\right)U = 0,$$

$$\frac{\mathrm{d}}{\mathrm{d}v}\left((1-v^2)\frac{\mathrm{d}V}{\mathrm{d}v}\right) + \left(-\kappa l^2 v^2 + \Lambda - \frac{m^2}{1-v^2}\right)V = 0, \qquad (2.94)$$

$$\frac{\mathrm{d}^2}{\mathrm{d}\varphi^2}\Phi + m^2\Phi = 0,$$

where m and Λ are constants of separation.

In terms of the variables $u = \cosh\xi$ and $v = \cos\eta$, Eq. (2.93) becomes

$$x = l\sinh\xi\sin\eta\cos\varphi, \qquad y = l\sinh\xi\sin\eta\sin\varphi,$$
$$z = l\cosh\xi\cos\eta, \qquad (\rho \equiv \sqrt{x^2+y^2} = l\sinh\xi\sin\eta). \qquad (2.95)$$

As is easily shown, the coordinate surfaces are

$$\frac{z^2}{(l\cosh\xi)^2} + \frac{x^2+y^2}{(l\sinh\xi)^2} = 1, \qquad \frac{z^2}{(l\cos\eta)^2} - \frac{x^2+y^2}{(l\sin\eta)^2} = 1,$$

so that the constant ξ surface gives an ellipsoid of revolution with a longer semi-axis in the z direction (see Fig. 2.4(a)), by which (ξ, η, φ) is called the **prolate spheroidal coordinate system**. If we regard ξ, η, φ as u_1, u_2, u_3, then we have (from Eqs. (2.65) and (2.62))

$$h_1 = h_2 = l\sqrt{\sinh^2\xi + \sin^2\eta}, \qquad h_3 = l\sinh\xi\sin\eta, \qquad (2.96)$$

and (from Eq. (2.70)),

$$\triangle\Psi = \frac{1}{l^2(\sinh^2\xi + \sin^2\eta)}\left[\frac{1}{\sinh\xi}\frac{\partial}{\partial\xi}\left(\sinh\xi\frac{\partial\Psi}{\partial\xi}\right) + \frac{1}{\sin\eta}\frac{\partial}{\partial\eta}\left(\sin\eta\frac{\partial\Psi}{\partial\eta}\right)\right]$$
$$+ \frac{1}{l^2\sinh^2\xi\sin^2\eta}\frac{\partial^2\Psi}{\partial\varphi^2}. \tag{2.97}$$

e. Oblate spheroidal coordinates (ξ, η, φ)

The replacement of $u \to iu$ and $il \to l$ in Eq. (2.93), yields the following oblate spheroidal coordinate system:

$$x = l\sqrt{(1+u^2)(1-v^2)}\cos\varphi,$$
$$y = l\sqrt{(1+u^2)(1-v^2)}\sin\varphi, \tag{2.98}$$
$$z = luv,$$
$$(0 \le u < \infty, \ |v| \le 1, \ 0 \le \varphi < 2\pi).$$

The surfaces of constant u give a family of confocal ellipsoids of revolution, those of constant v give a family of confocal hyperboloids of revolution of one sheets, and those of constant φ give the coaxial planes.

In terms of $u = \sinh\xi$ and $v = \cos\eta$, we have

$$x = l\cosh\xi\sin\eta\cos\varphi, \quad y = l\cosh\xi\sin\eta\sin\varphi,$$
$$z = l\sinh\xi\cos\eta, \quad (\rho = \sqrt{x^2+y^2} = l\cosh\xi\sin\eta). \tag{2.99}$$

It is also easily understood that surfaces of constant ξ describe ellipsoids of revolution with a shorter semi-axis in the z direction, by which (ξ, η, φ) is called the **oblate spheroidal coordinate system**. Figure 2.4(b) exemplifies the oblate spheroidal coordinate system, where the z axis is chosen on the q_2 axis, φ as the angle of rotation around the z axis, and the x axis on the q_1 axis in the meridian plane ($\varphi = 0$). In this case,

$$h_1 = h_2 = l\sqrt{\cosh^2\xi - \sin^2\eta}, \quad h_3 = l\cosh\xi\sin\eta, \tag{2.100}$$

so that we have

$$\triangle\Psi = \frac{1}{l^2(\cosh^2\xi - \sin^2\eta)}\left[\frac{1}{\cosh\xi}\frac{\partial}{\partial\xi}\left(\cosh\xi\frac{\partial\Psi}{\partial\xi}\right) + \frac{1}{\sin\eta}\frac{\partial}{\partial\eta}\left(\sin\eta\frac{\partial\Psi}{\partial\eta}\right)\right]$$
$$+ \frac{1}{l^2\cosh^2\xi\sin^2\eta}\frac{\partial^2\Psi}{\partial\varphi^2}. \tag{2.101}$$

f. Spherical coordinates (r, θ, φ)

Putting $u = r/l$ and $v = \cos\theta$ in Eq. (2.93), and taking the limit $l \to 0$ yield the **spherical coordinate system** (r, θ, φ):

$$x = r\sin\theta\cos\varphi, \quad y = r\sin\theta\sin\varphi, \quad z = r\cos\theta. \qquad (2.102)$$

(See Fig. 2.5(a).) The surface $r \equiv \sqrt{x^2 + y^2 + z^2} = $ constant gives the sphere centered at origin O. In this case,

$$h_1 = 1, \quad h_2 = r, \quad h_3 = r\sin\theta, \qquad (2.103)$$

and

$$\triangle\Psi = \frac{1}{r^2}\frac{\partial}{\partial r}\left(r^2\frac{\partial\Psi}{\partial r}\right) + \frac{1}{r^2}\left[\frac{1}{\sin\theta}\frac{\partial}{\partial\theta}\left(\sin\theta\frac{\partial\Psi}{\partial\theta}\right) + \frac{1}{\sin^2\theta}\frac{\partial^2\Psi}{\partial\varphi^2}\right]. \quad (2.104)$$

If the solution of the Helmholtz equation is sought in the spherical coordinate system of the form $\Psi = R(r)\Theta(\theta)\Phi(\varphi)$, then the separated equations depending on respective variables r, θ, φ are

$$\frac{1}{r^2}\frac{\mathrm{d}}{\mathrm{d}r}\left(r^2\frac{\mathrm{d}R}{\mathrm{d}r}\right) + \left(\kappa - \frac{\Lambda}{r^2}\right)R = 0,$$

$$\frac{1}{\sin\theta}\frac{\mathrm{d}}{\mathrm{d}\theta}\left(\sin\theta\frac{\mathrm{d}\Theta}{\mathrm{d}\theta}\right) + \left(\Lambda - \frac{m^2}{\sin^2\theta}\right)\Theta = 0, \qquad (2.105)$$

$$\frac{\mathrm{d}^2}{\mathrm{d}\varphi^2}\Phi + m^2\Phi = 0.$$

Note that the equation for Θ is rewritten by the use of $v = \cos\theta$:

$$\frac{\mathrm{d}}{\mathrm{d}v}\left[(1 - v^2)\frac{\mathrm{d}\Theta}{\mathrm{d}v}\right] + \left(\Lambda - \frac{m^2}{1 - v^2}\right)\Theta = 0. \qquad (2.106)$$

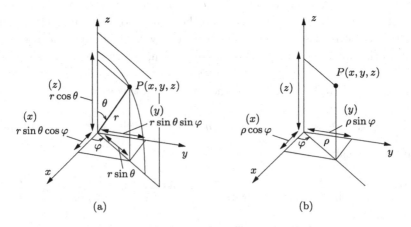

Fig. 2.5   (a) Spherical coordinates, and (b) cylindrical coordinates.

The solution for Θ of Eqs. (2.105) and (2.106) is given by the Legendre function (see §A.3). The spherical coordinate system is used for the problems including spherical boundaries, such as §§2.4.4c, §§2.6.1b, c, and §§2.6.4a.

g. Paraboloidal coordinates (ξ, η, φ)

By shifting the origin of z to l in Eq. (2.93) *i.e.*, $z = l(uv - 1)$, putting $u = \sqrt{1 + \xi^2/l}$, $v = \sqrt{1 - \eta^2/l}$ into x, y, z, and taking the limit $l \to \infty$, we obtain the **paraboloidal coordinate system** (ξ, η, φ):

$$x = \xi\eta\cos\varphi, \quad y = \xi\eta\sin\varphi, \quad z = \frac{1}{2}(\xi^2 - \eta^2), \quad (\rho \equiv \sqrt{x^2 + y^2} = \xi\eta).$$
(2.107)

Figure 2.6 shows the paraboloidal coordinate system, where we choose the q_2 axis as the z axis, the rotation angle around this axis as φ, and the q_1 axis as the x axis in the meridian plane $(\varphi = 0)$. Surfaces of constant ξ and those of constant η both give a family of paraboloids of revolution:

$$z = \frac{1}{2}\left(\xi^2 - \frac{x^2 + y^2}{\xi^2}\right), \quad z = \frac{1}{2}\left(\frac{x^2 + y^2}{\eta^2} - \eta^2\right).$$
(2.108)

In the present coordinate system,

$$h_1 = h_2 = \sqrt{\xi^2 + \eta^2}, \quad h_3 = \xi\eta,$$
(2.109)

from Eqs. (2.65) and (2.62), so that we have (from Eq. (2.70))

$$\triangle\Psi = \frac{1}{\xi^2 + \eta^2}\left[\frac{1}{\xi}\frac{\partial}{\partial\xi}\left(\xi\frac{\partial\Psi}{\partial\xi}\right) + \frac{1}{\eta}\frac{\partial}{\partial\eta}\left(\eta\frac{\partial\Psi}{\partial\eta}\right)\right] + \frac{1}{\xi^2\eta^2}\frac{\partial^2\Psi}{\partial\varphi^2}.$$
(2.110)

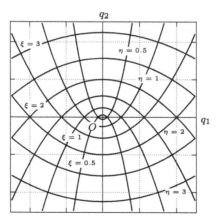

Fig. 2.6 Paraboloidal coordinates.

If the solution of the Helmholtz equation is sought in this coordinate system of the form $\Psi = \Xi(\xi)H(\eta)\Phi(\varphi)$, then the separated equations depending on respective variables ξ, η, φ are

$$\frac{1}{\xi}\frac{d}{d\xi}\left(\xi\frac{d\Xi}{d\xi}\right) + \left(\kappa\xi^2 - \frac{m^2}{\xi^2} + \lambda\right)\Xi = 0,$$

$$\frac{1}{\eta}\frac{d}{d\eta}\left(\eta\frac{dH}{d\eta}\right) + \left(\kappa\eta^2 - \frac{m^2}{\eta^2} - \lambda\right)H = 0, \tag{2.111}$$

whereas the equation for $\Phi(\varphi)$ is the same as the one in §§2.3.1d, f. Furthermore, if we put $t = \xi^2$ (or η^2), then $w = \Xi$ (or H) satisfies

$$\frac{d^2w}{dt^2} + \frac{1}{t}\frac{dw}{dt} + \left(\frac{\kappa}{4} \pm \frac{\lambda}{4t} - \frac{m^2}{4t^2}\right)w = 0, \tag{2.112}$$

where plus and minus signs correspond to $w = \Xi$ and $w = H$, respectively.

h. Cylindrical coordinates (ρ, φ, z)

By putting $u^2 = 1 + \rho^2/l^2, v = z/l$ in Eq. (2.93), and taking the limit $l \to \infty$, we have the circular cylindrical coordinate system, or simply the **"cylindrical coordinate system"** (ρ, φ, z):

$$x = \rho\cos\varphi, \quad y = \rho\sin\varphi, \quad z = z. \tag{2.113}$$

The surface $\rho \equiv \sqrt{x^2 + y^2} = $ constant gives a circular cylinder with the z axis passing through its center (see Fig. 2.5(b)). In this case,

$$h_1 = 1, \quad h_2 = \rho, \quad h_3 = 1, \tag{2.114}$$

and (from Eq. (2.70))

$$\Delta\Psi = \frac{1}{\rho}\frac{\partial}{\partial\rho}\left(\rho\frac{\partial\Psi}{\partial\rho}\right) + \frac{1}{\rho^2}\frac{\partial^2\Psi}{\partial\varphi^2} + \frac{\partial^2\Psi}{\partial z^2}. \tag{2.115}$$

If the solution of the Helmholtz equation is sought in the circular cylindrical coordinate system (ρ, φ, z) of the form $\Psi = R(\rho)\Phi(\varphi)Z(z)$, then the separated equations depending on respective variables ρ, φ, z are

$$\frac{1}{\rho}\frac{d}{d\rho}\left(\rho\frac{dR}{d\rho}\right) + \left(\kappa - \lambda - \frac{m^2}{\rho^2}\right)R = 0,$$

$$\frac{d^2}{d\varphi^2}\Phi + m^2\Phi = 0, \tag{2.116}$$

$$\frac{d^2Z}{dz^2} + \lambda Z = 0.$$

The solution of R in Eq. (2.116) is given by the Bessel function (see §§A.2). The cylindrical coordinates (ρ, φ, z) are useful for the problems

with circular cylindrical boundaries, which are shown in many examples
e.g. §§2.3.3a, §§2.4.4b, §§2.6.1–2.6.4, §§3.3.4–3.3.5, *etc.* Two-dimensional
polar coordinates (ρ, φ) are similar to the cylindrical coordinates in which
z dependence is omitted. Polar coordinates are suitable to solve the problems with circular boundaries, or to analyze the two-dimensional field that
depends on the distance from a certain point.

i. Elliptic cylinder coordinates (ξ, η, z)

We dealt with the general tri-axial spheroid in §§2.3.1d, in which semi-
axes are $\sqrt{u_1 - a_1} < \sqrt{u_1 - a_2} < \sqrt{u_1 - a_3}$. Here, we consider the limiting
case in which the largest semi-axis becomes infinite. To make $u_1 - a_3$ larger
than $u_1 - a_1$ and $u_1 - a_2$, while the latter two are of the same order, we
put $a_1 - a_2 = l^2$, $u_1 = a_2 + l^2 u^2$, $u_2 = a_2 + l^2 v^2$, $u_3 = a_3 + z^2$ in Eq. (2.80),
and take the limit $a_1, a_2 \to \infty$. Furthermore, to choose the x axis as the
longer radius, we exchange x and y. Then we have

$$x = luv, \quad y = l\sqrt{(u^2 - 1)(1 - v^2)}, \quad z = z, \qquad (2.117)$$

or in terms of $u = \cosh \xi$ and $v = \cos \eta$

$$x = l \cosh \xi \cos \eta, \quad y = l \sinh \xi \sin \eta, \quad z = z. \qquad (2.118)$$

Surfaces of constant u (or ξ) give a family of ellipses, whereas those of
constant v (or η) give a family of hyperbolas, by which (ξ, η, z) is called
the **elliptic cylinder coordinate system**. The coordinate curves in the
xy plane are the same as those of $q_2 q_1$ of the prolate spheroidal coordinate
system (Fig. 2.4(a)), which are taken perpendicular to the z axis.

Here, the Helmholtz equation for $U(u)$, which is a part of $\Psi = U(u)V(v)Z(z)$, is given by

$$\sqrt{u^2 - 1}\, \frac{d}{du}\left(\sqrt{u^2 - 1}\, \frac{dU}{du}\right) + \left[l^2(\kappa - \lambda)u^2 + \Lambda\right] U = 0,$$

and similarly to V and v.[7] In other words, Eq. (2.91) is given in terms of
$w = U$ (or V) and $\zeta = u$ (or v), as

$$\frac{d^2 w}{d\zeta^2} + \frac{1}{2}\left(\frac{1}{\zeta - 1} + \frac{1}{\zeta + 1}\right)\frac{dw}{d\zeta} + \frac{l^2(\kappa - \lambda)\zeta^2 + \Lambda}{\zeta^2 - 1} w = 0. \qquad (2.119)$$

---

[7]For $w = V(v)$ and $\zeta = v$,

$$\sqrt{1 - v^2}\, \frac{d}{dv}\left(\sqrt{1 - v^2}\, \frac{dV}{dv}\right) - \left[l^2(\kappa - \lambda)v^2 + \Lambda\right] V = 0.$$

In this case, the scale factors are

$$h_1 = h_2 = l\sqrt{\cosh^2 \xi - \cos^2 \eta}, \quad h_3 = 1, \qquad (2.120)$$

and

$$\triangle\Psi = \frac{1}{l^2(\cosh^2 \xi - \cos^2 \eta)} \left(\frac{\partial^2 \Psi}{\partial \xi^2} + \frac{\partial^2 \Psi}{\partial \eta^2} \right) + \frac{\partial^2 \Psi}{\partial z^2}. \qquad (2.121)$$

If the solution of the Helmholtz equation is sought in this coordinate system (ξ, η, z) of the form $\Psi = \Xi(\xi)H(\eta)Z(z)$, then the separated equations depending on respective variables are

$$\frac{d^2\Xi}{d\xi^2} + [l^2(\kappa - \lambda)\cosh^2 \xi + \Lambda]\Xi = 0,$$

$$\frac{d^2 H}{d\eta^2} - [l^2(\kappa - \lambda)\cos^2 \eta + \Lambda]H = 0, \qquad (2.122)$$

$$\frac{d^2 Z}{dz^2} + \lambda Z = 0.$$

j. Parabolic cylinder coordinates (ξ, η, z)

If we shift the origin of x to l in §§2.3.1i, and make similar changes as in the process shown in §§2.3.1g, we have

$$x = \frac{1}{2}(\xi^2 - \eta^2), \quad y = \xi\eta, \quad z = z. \qquad (2.123)$$

The coordinate curves in the xy plane are the same as those of q_2q_1 of the paraboloid of revolution in the meridian plane (Fig. 2.6), which are taken perpendicular to the z axis, by which (ξ, η, z) is called the **parabolic cylinder coordinate system**. In this case,

$$h_1 = h_2 = \sqrt{\xi^2 + \eta^2}, \quad h_3 = 1, \qquad (2.124)$$

and

$$\triangle\Psi = \frac{1}{\xi^2 + \eta^2} \left(\frac{\partial^2 \Psi}{\partial \xi^2} + \frac{\partial^2 \Psi}{\partial \eta^2} \right) + \frac{\partial^2 \Psi}{\partial z^2}. \qquad (2.125)$$

If the solution of the Helmholtz equation is sought in this coordinate system (ξ, η, z) of the form $\Psi = \Xi(\xi)H(\eta)Z(z)$, then the separated equation for $w = \Xi(\xi)$ or $H(\eta)$ is

$$\frac{d^2w}{d\zeta^2} + [(\kappa - \lambda)\zeta^2 \pm \Lambda]w = 0, \qquad (2.126)$$

where $\zeta = \xi$ and η correspond to $w = \Xi(\xi)$ and $H(\eta)$, respectively. Furthermore, if we put $t = \zeta^2$, then we have

$$\frac{d^2w}{dt^2} + \frac{1}{2t}\frac{dw}{dt} + \left(\frac{\kappa - \lambda}{4} \pm \frac{\Lambda}{4t} \right)w = 0. \qquad (2.127)$$

The equation for $Z(z)$ is the same as the one previously shown at §§2.3.1**h**, **i**.

In the following, we shall briefly show the definitions and their coordinate curves, which are basically the same, but are differently named stemmed from their practical application.

k. Bispherical coordinates (ξ, η, φ),
Bipolar cylinder coordinates (ξ, η, z)

Bipolar coordinates (ξ, η) in a two-dimensional space are defined by

$$q_1 = c\frac{\sinh\eta}{\cosh\eta - \cos\xi}, \qquad q_2 = c\frac{\sin\xi}{\cosh\eta - \cos\xi}, \qquad (c > 0). \qquad (2.128)$$

Coordinate curves of this system are shown in Fig. 2.7(a). If we choose the q_2 axis as the axis of rotation, around which the angle of rotation φ is taken, they form a **bispherical coordinate system** (ξ, η, φ). Surfaces of constant η give a family of eccentric spheres. But, if we choose the coordinates (q_1, q_2) in the plane perpendicular to the z axis, they constitute the **bipolar cylinder coordinate system** (ξ, η, z).

These coordinate systems are often applied to analyze the problems with boundaries consisting of two spheres or two parallel circular cylinders, including a sphere and plane or a circular cylinder parallel to a plane as special cases.

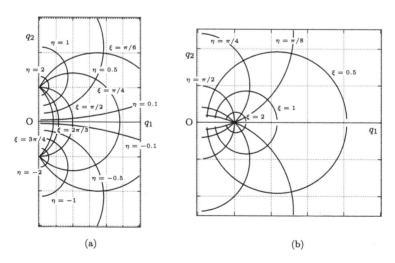

(a)   (b)

Fig. 2.7   (a) Bispherical coordinates, and (b) toroidal coordinates.

1. Toroidal coordinates (ξ, η, φ)

The coordinate system, in which ξ and η, as well as q_1 and q_2 are exchanged in Eq. (2.128):

$$q_1 = c\frac{\sin\eta}{\cosh\xi - \cos\eta}, \qquad q_2 = c\frac{\sinh\xi}{\cosh\xi - \cos\eta}, \qquad (c > 0) \qquad (2.129)$$

is shown in Fig. 2.7(b). If the angle of rotation φ is taken around the q_2 axis, surfaces of constant ξ give a family of toroids, or doughnuts. This coordinate system (ξ, η, φ) is called the **toroidal coordinate system**. This coordinate system is often applied to analyze the problems with a doughnut-shaped boundary (a circular ring with a finite thickness), the boundary with cusped region, *etc.*

If the shape of the boundary coincides with one of the coordinate surfaces, the use of a relevant coordinate system will sometimes be suitable for an analytic calculation. For further details on the orthogonal curvilinear coordinate system, see *e.g.*, [Morse and Feshbach (1953)](Chap. 5).

2.3.2 *Eigenvalues and eigenfunctions in the Sturm–Liouville equations*

In general, a second-order linear homogeneous ODE:

$$A(x)\frac{d^2u}{dx^2} + B(x)\frac{du}{dx} + C(x)u = 0 \qquad (2.130)$$

is reduced to

$$L[u] + \lambda r(x)u = 0, \qquad (2.131)$$

by multiplying

$$H(x) = \frac{1}{A(x)}\exp\left[\int^x \frac{B(\xi)}{A(\xi)}\,d\xi\right], \qquad (2.132)$$

where $A(x)$ is assumed to be non-zero at any point in the interval under consideration. In the above expression, operator L is defined by

$$L[u] \equiv \frac{d}{dx}\left[p(x)\frac{du}{dx}\right] + q(x)u, \qquad (2.133)$$

where $p(x) > 0$, and we can assume $r(x) > 0$ by the suitable choice of $q(x)$.[8] An ordinary differential equation of the form Eq. (2.131) with (2.133) is

---

[8]In the above calculation, we put

$$p(x) = \exp\left[\int^x \frac{B(\xi)}{A(\xi)}\,d\xi\right], \qquad q(x) + \lambda r(x) = \frac{C(x)}{A(x)}\exp\left[\int^x \frac{B(\xi)}{A(\xi)}\,d\xi\right].$$

called the **Sturm–Liouville equation**. As will be shown later, imposing the boundary conditions or other conditions allows only the specified values of λ, which are called the **eigenvalues**, and the solution obtained for a specific eigenvalue is called the **eigenfunction** associated with that eigenvalue.

For simplicity, we assume that the independent variable x varies in a finite interval $[a, b]$, in which $p(x)$, $q(x)$ and $r(x)$ are continuous, and $p(x) > 0$, $r(x) > 0$. The boundary conditions at both ends of the interval are assumed homogeneous and are given by

$$p(a)u'(a)\sin\alpha - u(a)\cos\alpha = 0,$$
$$p(b)u'(b)\sin\beta - u(b)\cos\beta = 0. \tag{2.134}$$

The following particular cases should be mentioned.

(1) $\alpha = \beta = 0$, which implies $u(a) = u(b) = 0$: This type of condition is called the **Dirichlet condition**, or homogeneous condition of the first kind.

(2) $\alpha = \beta = \pi/2$, which implies $u'(a) = u'(b) = 0$: This type of condition is called the **Neumann condition**, or homogeneous condition of the second kind.

(3) Otherwise: In this case, the condition at $x = a$ is given by $u'(a) - h(a)u(a) = 0$, $h(a) = \cos\alpha/p(a)\sin\alpha$, (and similarly at $x = b$), so that the condition is specified by the combination of the value of the function and the differential coefficient. This type of condition is called the **Cauchy condition**, or homogeneous condition of the third kind.

$\triangleleft$

a. Orthogonality of eigenfunctions

Let the functions u_n and u_m be the solutions of Eq. (2.131), which are associated with the different eigenvalues $\lambda = \lambda_n$ and λ_m, respectively. Then they satisfy

$$\frac{\mathrm{d}}{\mathrm{d}x}\left(p\frac{\mathrm{d}u_n}{\mathrm{d}x}\right) + (q + \lambda_n r)u_n = 0,$$

$$\frac{\mathrm{d}}{\mathrm{d}x}\left(p\frac{\mathrm{d}u_m}{\mathrm{d}x}\right) + (q + \lambda_m r)u_m = 0.$$

If we multiply the first and second equations above by u_m and u_n, respectively, and subtract them, we have

$$u_m\frac{\mathrm{d}}{\mathrm{d}x}\left(p\frac{\mathrm{d}u_n}{\mathrm{d}x}\right) - u_n\frac{\mathrm{d}}{\mathrm{d}x}\left(p\frac{\mathrm{d}u_m}{\mathrm{d}x}\right) + (\lambda_n - \lambda_m)ru_nu_m = 0. \tag{2.135}$$

Integration of the latter in the interval $[a, b]$ yields

$$(\lambda_m - \lambda_n) \int_a^b r u_n u_m \, \mathrm{d}x = \left[p \left(u_m \frac{\mathrm{d}u_n}{\mathrm{d}x} - u_n \frac{\mathrm{d}u_m}{\mathrm{d}x} \right) \right]_a^b = 0, \qquad (2.136)$$

where the homogeneous boundary conditions (of any type of the above-mentioned conditions) at $x = a, b$ are used. This relation shows that eigenfunctions belonging to different eigenvalues are orthogonal to each other in the interval $[a, b]$ with a weight function $r(x)$.

b. Zeros of the eigenfunctions

Consider the sequence of real eigenvalues $\lambda_1 < \lambda_2 < ... < \lambda_n < ...$, and the associated eigenfunctions $u_1 < u_2 < ... < u_n < ...$ obtained under the homogeneous boundary conditions. We consider, for instance, the case $n = 1, 2$ and integrate Eq. (2.135) over the interval $[a, \xi]$. The integrated value at $x = a$ is 0 owing to the boundary conditions (2.134) for u_1 and u_2 at $x = a$. To be more specific, we assume that $x = \xi$ is a zero of u_1 (see Fig. 2.8), then we have

$$(\lambda_2 - \lambda_1) \int_a^\xi r u_1 u_2 \, \mathrm{d}x = \left[p \left(u_2 \frac{\mathrm{d}u_1}{\mathrm{d}x} \right) \right]_{x=\xi}. \qquad (2.137)$$

Here, we assume, without loss of generality, that u_1 is positive at $a < x < \xi$, and becomes negative at $x > \xi$, and hence $\mathrm{d}u_1/\mathrm{d}x < 0$ at $x = \xi$.

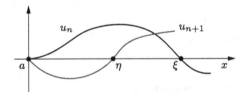

Fig. 2.8   Zeros of eigenfunctions.

If u_2 has a definite sign, *e.g.*, if $u_2 > 0$ in the interval $[a, \xi]$, then the l.h.s. of the above integral is positive whereas the r.h.s. is negative (since $p(x) > 0$, $r(x) > 0$ have been assumed), which leads to a contradiction (conversely, if $u_2 < 0$ in this interval, it also leads to a contradiction). Accordingly, we conclude that u_2 must change signs in the interval $a < x < \xi$, *i.e.*, u_2 must have a zero in this interval. The repetition of the above argument shows in general that zeros of u_{n+1} appear between successive zeros of u_n.

The number of zeros depends on the boundary conditions. For example, in the vibration of a string with both ends fixed, the lowest eigenfunction u_1

has two zeros (at both ends), so that the n-th eigenfunction $u_n(x)$ has $n+1$ zeros. Meanwhile, as observed in the vibration of an elastic plate in which one of the ends is fixed whereas the other is free, the lowest eigenfunction u_1 has only one zero, so that the n-th eigenfunction $u_n(x)$ has n zeros. For more details, see *e.g.*, [Morse and Feshbach (1953)](Chap. 6). In any case, eigenfunctions $u_n(x)$ associated with λ_n are created in the interval $[a, b]$ with an increase in n, so that an infinite series of functions is created.

c. Eigenfunction expansion

As shown in the previous subsection, we can create an infinite series of mutually orthogonal eigenfunctions $u_n(x)$, ($n = 1, 2, 3, ...$) in the interval $[a, b]$. By the use of these eigenfunctions, we can expand a piecewise smooth function $f(x)$ defined in the same interval as

$$f(x) = \sum_{n=1}^{\infty} c_n u_n(x), \qquad (2.138)$$

$$c_n = \frac{(f, u_n)}{(u_n, u_n)}, \qquad (2.139)$$

where

$$(u_n, u_m) = \int_a^b r(\xi)\, u_n(\xi)\, u_m(\xi)\, \mathrm{d}\xi. \qquad (2.140)$$

If the series of functions $u_n(x)$ is a complete system,[9] and the infinite series $\sum_{n=1}^{\infty} c_n u_n(x)$ converges uniformly, Eq. (2.138) becomes useful. Such an infinite series is called the **Fourier-type series** with respect to the complete orthogonal function system $\{u_n(x)\}$. We show some examples of the Sturm-Liouville equations in Table 2.1.

2.3.3 *Eigenfunction expansions*

In terms of the series of functions $\phi_n(x)$ that satisfy the given ODE and boundary conditions, we can create the orthonormal function family. To

---

[9]When an arbitrary continuous function is expanded in a series of orthogonal functions $u_n(x)$, and the integral over the defined interval of "the square of the difference between the given function and the series expansion up to N terms" tends to zero with an increase in N:

$$\lim_{N \to \infty} \int_a^b \left(f(x) - \sum_{n=1}^{N} c_n u_n(x) \right)^2 r(x)\, \mathrm{d}x = 0,$$

then the function series $u_n(x)$ forms a complete system. This criterion is called the **convergence in the mean**.

Table 2.1 Examples of the Sturm–Liouville equations.

ODE	$p(x)$	$q(x)$	$r(x)$	λ	interval
(a) Legendre $P_n(x) \equiv P_n^0(x)$	$1 - x^2$	0	1	$n(n+1)$	$[-1, 1]$
associated Legendre $P_n^m(x)$	$1 - x^2$	$-\dfrac{m^2}{1 - x^2}$	1	$n(n+1)$	$[-1, 1]$
(b) Chebyshev $T_n(x)$	$\sqrt{1 - x^2}$	0	$\dfrac{1}{\sqrt{1 - x^2}}$	n^2	$[-1, 1]$
(c) Bessel $J_n(j_{nm}x)$	x	$-\dfrac{n^2}{x}$	x	j_{nm}^2	$[0, 1]$
(d) Laguerre $L_n(x) \equiv L_n^0(x)$	xe^{-x}	0	e^{-x}	n	$[0, \infty)$
associated Laguerre $L_n^m(x)$	$x^{m+1}e^{-x}$	0	$x^m e^{-x}$	n	$[0, \infty)$
(e) Hermite $H_n(x)$	e^{-x^2}	0	e^{-x^2}	$2n$	$(-\infty, \infty)$
(f) simple harmonic osc.	1	0	1	n^2	$[0, 2\pi]$

(a) $(1 - x^2)u'' - 2xu' + \left[n(n+1) - m^2/(1 - x^2)\right]u = 0 \rightarrow u = P_n^m(x)$
(b) $(1 - x^2)u'' - xu' + n^2u = 0 \rightarrow u = T_n(x)$
(c) $x^2u'' + xu' + (j_{nm}^2 x^2 - n^2)u = 0 \rightarrow u = J_n(j_{nm}x)$
 (where j_{nm} is the m-th zero of $J_n(x)$, see also §§2.3.4)
(d) $xu'' + (m + 1 - x)u' + nu = 0 \rightarrow u = L_n^m(x)$
(e) $u'' - 2xu' + 2nu = 0 \rightarrow u = H_n(x)$
(f) $u'' + n^2u = 0 \rightarrow u = \{\sin(nx), \cos(nx)\}$

do so, we first define the integral, in which the product of the two functions is integrated over the interval $[a, b]$ with a weight function $r(x)$:

$$(\phi_m, \phi_n) = \int_a^b r(\xi)\phi_m(\xi)\phi_n(\xi)\, d\xi. \tag{2.141}$$

To obtain a series of mutually orthogonal functions, we create a set of functions $\{\phi_n(x), \ n = 1, 2, 3, ...\}$ such that they satisfy the condition $(\phi_m, \phi_n) \propto \delta_{mn}$.[10] Furthermore, we calculate

$$(\phi_m, \phi_m) = \int_a^b r(\xi)\phi_m(\xi)^2\, d\xi = N_m, \tag{2.142}$$

and re-define new functions by $\tilde{\phi}_m = \phi_m/\sqrt{N_m}$, so that they constitute a set of orthonormal functions:

$$(\tilde{\phi}_m, \tilde{\phi}_n) = \delta_{mn}. \tag{2.143}$$

---

[10]The symbol δ_{mn} is called the Kronecker's delta, and is equal to 1 for $m = n$, and 0 otherwise.

Using the latter orthonormal function system, we can expand $f(x)$ as follows:

$$f(x) = \sum_{n=1}^{\infty} c_n \tilde{\phi}_n(x), \quad c_m = (f(x), \tilde{\phi}_m(x)), \quad (2.144)$$

which is called an **eigenfunction expansion**. Hereafter, we omit the "tilde" (a symbol of accent over the letter) in the expression $\tilde{\phi}_m$, and regard ϕ_m as the normalized orthogonal function. In terms of the latter expansion, with a weight function $r(x) > 0$, the relation

$$\int_a^b \left(f(x) - \sum_{n=1}^{\infty} c_n \phi_n(x) \right)^2 r(x)\, \mathrm{d}x \geq 0 \quad (2.145)$$

is always satisfied, where the equality holds only if the function system is complete. If the above equation is expanded, we have

$$\int_a^b f(x)^2 r(x)\, \mathrm{d}x \geq \sum_{n=1}^{\infty} c_n^2, \quad (2.146)$$

which is called the **Bessel inequality**. ◁

In the following, we show some examples of eigenfunction expansions.

a. Eigenfunction expansion by trigonometric functions

A simple harmonic oscillation is a typical phenomenon observed in physics as the motion of a body attached to a spring, which moves back and forth from the equilibrium position owing to the restoring force. The motion obeys the Newton's second law

$$m \frac{\mathrm{d}^2 x}{\mathrm{d}t^2} = -kx,$$

where m is the mass of the body, $k (> 0)$ is the spring constant of the restoring force, x is the displacement of the body from the equilibrium position, and t is the time. This type of equation appears in many fields of science and engineering, irrespective of the time variation or spatial variation of the pertaining dependent variables. In the latter, it takes the form

$$\frac{\mathrm{d}^2 u}{\mathrm{d}x^2} + \lambda u = 0 \quad (\lambda > 0), \quad (2.147)$$

and the solution is given by the eigenfunctions that depend on the boundary conditions, *i.e.*,

(i) eigenfunctions with both ends $x = 0$, l fixed:

$$\sqrt{\frac{2}{l}} \sin\left(\frac{n\pi}{l}x\right), \quad (n = 1, 2, 3, ...), \tag{2.148}$$

(ii) eigenfunctions with both ends $x = 0$, l free:

$$\sqrt{\frac{2}{l}} \cos\left(\frac{n\pi}{l}x\right), \quad (n = 1, 2, 3, ...), \tag{2.149}$$

(iii) eigenfunctions with periodic boundary conditions $u(-l) = u(l)$, and $u'(-l) = u'(l)$:

$$\frac{1}{\sqrt{2l}} \exp\left(\frac{n\pi i}{l}x\right), \quad (n = 0, \pm1, \pm2, ...). \tag{2.150}$$

All of these eigenfunction series are orthonormal and are used in the Fourier-type series expansion. They all have common eigenvalues $(n\pi/l)^2$. Using such an eigenfunction series $\{u_n(x)\}$, the displacement of the oscillator is generally expressed as

$$u(x) = \sum_{n=1}^{\infty} c_n u_n(x), \quad c_m = \big(u(x), u_m(x)\big). \tag{2.151}$$

◁

We show some examples of solving PDEs by the use of an eigenfunction expansion.

Example 2.3 (Two-dimensional potential problem).

A hollow metallic cylinder of radius a is divided into two sections through its axis, to which the electrostatic potentials $\pm V_0$ are given respectively. Obtain the electrostatic potential distribution V.

In this case, it is appropriate to use the cylindrical coordinate system (ρ, φ, z) with the z axis on the cylinder axis. Since the cylinder is infinitely long with a constant radius, there is no dependence on z, so that the equation for $V(\rho, \varphi)$ is given by the two-dimensional Laplace equation:

$$\frac{1}{\rho}\frac{\partial}{\partial \rho}\left(\rho\frac{\partial V}{\partial \rho}\right) + \frac{1}{\rho^2}\frac{\partial^2 V}{\partial \varphi^2} = 0, \tag{2.152}$$

and the boundary conditions are (with a suitable choice of direction $\varphi = 0$)

$$V(a, \varphi) = V_0 \quad (0 < \varphi < \pi), \qquad V(a, \varphi) = -V_0 \quad (-\pi < \varphi < 0). \tag{2.153}$$

Considering the geometric configuration and the boundary conditions, the potential V should be a periodic function of φ with period 2π, and that it is an odd function of φ, which allows us to expand V in the Fourier sine series:

$$V = \sum_{n=1}^{\infty} b_n(\rho) \sin(n\varphi).$$

Substitution of the above expansion into Eq. (2.152) yields

$$\frac{1}{\rho} \frac{d}{d\rho} \left(\rho \frac{d b_n}{d\rho} \right) - \frac{n^2}{\rho^2} b_n = 0.$$

The latter is an Euler-type ODE, so that it has the solution ρ^n or ρ^{-n}. Taking into account that the potential is finite in the region $\rho < a$, the solution ρ^{-n} is inappropriate, so that the allowable solution is

$$V = \sum_{n=1}^{\infty} c_n \rho^n \sin(n\varphi).$$

By applying the boundary condition (2.153), the coefficient c_n is determined to be $c_n = 4V_0/(\pi a^n n)$ (n: odd). Consequently, the potential is given by

$$V(\rho, \varphi) = \frac{4V_0}{\pi} \sum_{m=1}^{\infty} \frac{1}{2m-1} \left(\frac{\rho}{a} \right)^{2m-1} \sin[(2m-1)\varphi]. \tag{2.154}$$

◁

Example 2.4 (One-dimensional wave equation).

Obtain the solution of the wave equation (the same as Eq. (2.2) derived in §§2.1.1)

$$\frac{\partial^2 u}{\partial x^2} = \frac{1}{c^2} \frac{\partial^2 u}{\partial t^2}, \tag{2.155}$$

under the boundary conditions

$$u(0,t) = u(L,t) = 0, \tag{2.156}$$

and the initial conditions

$$u(x,0) = u_0(x), \qquad \frac{\partial u}{\partial t}(x,0) = v_0(x). \tag{2.157}$$

The solution u under consideration is continuous in $0 \le x \le L$, and satisfies the both-ends-fixed conditions, so that it should be written as

$$u(x,t) = \sum_{n=1}^{\infty} b_n(t) \sin\left(\frac{n\pi x}{L} \right). \tag{2.158}$$

To fulfill the governing equation, we must have

$$\frac{1}{c^2}\frac{d^2 b_n}{dt^2} = -\left(\frac{n\pi}{L}\right)^2 b_n, \tag{2.159}$$

where the orthogonality of the trigonometric functions has been considered. By solving the above ODE, we have the general solution

$$u(x,t) = \sum_{n=1}^{\infty}\left[A_n \cos\left(\frac{n\pi ct}{L}\right) + B_n \sin\left(\frac{n\pi ct}{L}\right)\right]\sin\left(\frac{n\pi x}{L}\right). \tag{2.160}$$

The coefficients A_n, B_n are determined by the initial conditions:

$$\sum_{n=1}^{\infty} A_n \sin\left(\frac{n\pi x}{L}\right) = u_0(x) \rightarrow A_n = \frac{2}{L}\int_0^L u_0(x)\sin\left(\frac{n\pi x}{L}\right)dx, \tag{2.161}$$

$$\sum_{n=1}^{\infty}\frac{n\pi c}{L}B_n \sin\left(\frac{n\pi x}{L}\right) = v_0(x) \rightarrow B_n = \frac{2}{n\pi c}\int_0^L v_0(x)\sin\left(\frac{n\pi x}{L}\right)dx. \tag{2.162}$$

The method of solution developed here is essentially the same as the finite integral transforms that are described later (see §3.2). Note that the method of separation of variables is often used for obtaining the stationary wave, which leads to the same results.[11] ◁

b. Eigenfunction expansions by polynomials

The Legendre differential equation

$$\frac{d}{dx}\left((1-x^2)\frac{du}{dx}\right) + \lambda u = 0 \tag{2.163}$$

---

[11]The method of separation of variables adopts the solution of the form $u(x,t) = X(x)T(t)$, which is substituted for the wave equation, and

$$\frac{X''(x)}{X(x)} = \frac{1}{c^2}\frac{T''(t)}{T(t)} = -\lambda^2 \text{ (constant)}$$

is solved:

$$X(x) = A\sin(\lambda x) + B\cos(\lambda x), \qquad T(t) = C\sin(c\lambda t) + D\cos(c\lambda t).$$

If the boundary condition $u = 0$ at $x = 0$ and L is imposed, then $B = 0$, and the condition $\sin(\lambda L) = 0$ requires $\lambda = n\pi/L$ (where n is an integer), which determines the eigenvalue λ. Consequently, the solution is given by the following expression (with a suitable choice of arbitrary constants)

$$u(x,t) = \sum_{n=1}^{\infty}\sin\left(\frac{n\pi}{L}x\right)\left[a_n \sin\left(\frac{nc\pi}{L}t\right) + b_n \cos\left(\frac{nc\pi}{L}t\right)\right],$$

which is of the same form as Eq. (2.160). The remaining constants a_n, b_n are determined by the initial conditions.

has solutions in terms of polynomials in the interval $[-1, 1]$. The eigenvalues are $\lambda = n(n + 1)$, and the orthonormal eigenfunctions are $\sqrt{n + \frac{1}{2}} P_n(x)$, the first few terms of which are

$$\sqrt{\frac{1}{2}}, \quad \sqrt{\frac{3}{2}} x, \quad \sqrt{\frac{5}{8}} (3x^2 - 1), \quad \sqrt{\frac{7}{8}} (5x^3 - 3x), \quad \cdots. \tag{2.164}$$

These functions are called the **Legendre polynomials** (see *e.g.*, Appendix A.3 of this textbook). ◁

If we pay attention to the expression of the above series that includes an increasing order (or power) of polynomial of x, we can anticipate to generate an **orthonormal polynomial** series, which, starting from a lower order of x to higher order polynomials, successively satisfies the orthogonality to the previous polynomials as well as the normalization. We shall perform the above procedures. First, for the zero-th order of x, we choose a constant $y_0 = c$, which is normalized as follows:

$$(y_0, y_0) = \int_{-1}^{1} c^2 \mathrm{d}x = 2c^2 = 1 \to c = \frac{1}{\sqrt{2}} \to y_0 = \frac{1}{\sqrt{2}}.$$

Second, we attempt $y_1 = bx$ by taking into account that the first order of x is independent of y_0. The orthogonality to y_0 is evident. By the normalization, y_1 is determined:

$$(y_1, y_1) = \int_{-1}^{1} b^2 x^2 \mathrm{d}x = \frac{2b^2}{3} = 1 \to b = \sqrt{\frac{3}{2}} \to y_1 = \sqrt{\frac{3}{2}} x.$$

Third, we consider $y_2 = ax^2 + bx + c$ in view of the requirement that it includes a next higher order of x that is independent of the previous ones. The orthogonality requirement gives

$$(y_2, y_0) = \frac{1}{\sqrt{2}} \int_{-1}^{1} (ax^2 + bx + c) \, \mathrm{d}x = \sqrt{2} \left(\frac{a}{3} + c \right) = 0,$$

$$(y_2, y_1) = \sqrt{\frac{3}{2}} \int_{-1}^{1} (ax^3 + bx^2 + cx) \, \mathrm{d}x = \frac{\sqrt{6}}{3} b = 0,$$

which yields $y_2 = a(x^2 - 1/3)$. The normalization requirement determines a, so that the third function is obtained:

$$(y_2, y_2) = a^2 \int_{-1}^{1} \left(x^2 - \frac{1}{3} \right)^2 \mathrm{d}x = \frac{8a^2}{45} = 1 \to a = 3\sqrt{\frac{5}{8}} \to y_2 = \sqrt{\frac{5}{8}} (3x^2 - 1),$$

which agrees with Eq. (2.164).

As shown in the above example, the approach to generate a function series successively that satisfy the orthogonality and normalization is called the **Gram–Schmidt orthogonalization method**. ◁

The orthogonality depends on the interval considered, as well as the weight function $r(x)$ adopted in the definition (2.141). In the following, we show some general expressions of the system of orthogonal polynomial functions that are frequently used in science and engineering, although the normalization is not necessarily made here.

(i) The choices of the interval $[-1, 1]$ and $r(x) = 1$ give the Legendre polynomials $P_n(x)$, whose first few terms are

$$P_0 = 1, \ P_1 = x, \ P_2 = \frac{1}{2}(3x^2 - 1), \ \cdots,$$

and in general

$$P_n(x) = \frac{1}{2^n n!}\frac{d^n}{dx^n}(x^2 - 1)^n, \quad \text{or} \quad \frac{(-1)^n}{2^n n!}\frac{d^n}{dx^n}(1 - x^2)^n.$$

The orthogonality relation is

$$\int_{-1}^{1} P_m(x)P_n(x)\,dx = \frac{2}{2n+1}\delta_{mn},$$

and the ODE that is satisfied by $y = P_n(x)$ is

$$(1 - x^2)y'' - 2xy' + n(n+1)y = 0.$$

(ii) The choices of the interval $(-\infty, \infty)$ and $r(x) = \exp(-x^2)$ give the **Hermite polynomials** $H_n(x)$, whose first few terms are

$$H_0 = 1, \ H_1 = 2x, \ H_2 = 4x^2 - 2, \ \cdots,$$

and in general

$$H_n(x) = (-1)^n e^{x^2}\frac{d^n}{dx^n}e^{-x^2}.$$

The orthogonality relation is

$$\int_{-\infty}^{\infty} e^{-x^2} H_m(x)H_n(x)\,dx = 2^n n!\sqrt{\pi}\,\delta_{mn},$$

and the ODE that is satisfied by $y = H_n(x)$ is

$$y'' - 2xy' + 2ny = 0.$$

Another definition of the Hermite polynomials is also used.[12]

---

[12]If the weight function is chosen as $r(x) = \exp(-x^2/2)$, it follows that

$$H_0 = 1, \ H_1 = x, \ H_2 = x^2 - 1, \ \cdots,$$

and in general

$$H_n(x) = (-1)^n e^{x^2/2}\frac{d^n}{dx^n}e^{-x^2/2}.$$

The orthogonality relation is

$$\int_{-\infty}^{\infty} e^{-x^2/2} H_m(x)H_n(x)\,dx = n!\sqrt{2\pi}\,\delta_{mn},$$

and the ODE that is satisfied by $y = H_n(x)$ is

$$y'' - xy' + ny = 0.$$

(iii) The choices of the interval $[0, \infty)$ and $r(x) = \exp(-x)$ give the **Laguerre polynomials** $L_n(x)$, whose first few terms are

$$L_0 = 1, \ L_1 = 1 - x, \ L_2 = 2 - 4x + x^2, \ \cdots,$$

and in general

$$L_n(x) = e^x \frac{d^n}{dx^n} (x^n e^{-x}).$$

The orthogonality relation is

$$\int_0^\infty e^{-x} L_m(x) L_n(x) \ dx = (n!)^2 \delta_{mn},$$

and the ODE that is satisfied by $y = L_n(x)$ is

$$xy'' + (1 - x)y' + ny = 0.$$

For further examples, see *e.g.*, [Morse and Feshbach (1953)](Chap. 6), [Abramowitz and Stegun (1964)](Chap. 22), [Arfken (1985)](Chap. 13).

2.3.4 *Fourier–Bessel Expansions*

Laplace equation $\triangle \Phi = 0$ is given in the cylindrical coordinate system (ρ, φ, z) as

$$\frac{1}{\rho} \frac{\partial}{\partial \rho} \left(\rho \frac{\partial \Phi}{\partial \rho} \right) + \frac{1}{\rho^2} \frac{\partial^2 \Phi}{\partial \varphi^2} + \frac{\partial^2 \Phi}{\partial z^2} = 0. \tag{2.165}$$

If the solution is sought by the separation-of-variable type $\Phi(\rho, \varphi, z) = f(\rho) \exp(in\varphi \pm kz)$, the equation satisfied by f is

$$\frac{d^2 f}{d\rho^2} + \frac{1}{\rho} \frac{df}{d\rho} + \left(k^2 - \frac{n^2}{\rho^2} \right) f = 0, \tag{2.166}$$

or using $k\rho = x$,

$$\frac{d^2 f}{dx^2} + \frac{1}{x} \frac{df}{dx} + \left(1 - \frac{n^2}{x^2} \right) f = 0. \tag{2.167}$$

The solution of the latter is given by $J_n(x)$, which is called the n-th order **Bessel function** of the first kind. Thus far we have regarded n as an integer. Hereafter, however, we extend n to a real number ν, and consider in general

$$\frac{d^2 y}{dx^2} + \frac{1}{x} \frac{dy}{dx} + \left(k^2 - \frac{\nu^2}{x^2} \right) y = \frac{1}{x} \frac{d}{dx} \left(x \frac{dy}{dx} \right) + \left(k^2 - \frac{\nu^2}{x^2} \right) y = 0, \tag{2.168}$$

where k is assumed a positive real number. The solution is given by the Bessel function $J_\nu(kx)$. As shown in Appendix A.2.1a, $J_{-\nu}(x)$ is independent of $J_\nu(x)$ for non-integer ν, whereas it is not independent if ν is an integer. In the latter, the **Neumann function** is introduced as a function independent of $J_\nu(x)$. The behaviors and properties of these functions are given in §§A.2.1.

Using these functions, we can draw some general features. We first notice that Eq. (2.168) multiplied by x belongs to the Sturm-Liouville-type ODE:

$$\frac{\mathrm{d}}{\mathrm{d}x}\left(x\frac{\mathrm{d}J_\nu(kx)}{\mathrm{d}x}\right) + \left(k^2 x - \frac{\nu^2}{x}\right)J_\nu(kx) = 0,$$

and also for a different parameter l:

$$\frac{\mathrm{d}}{\mathrm{d}x}\left(x\frac{\mathrm{d}J_\nu(lx)}{\mathrm{d}x}\right) + \left(l^2 x - \frac{\nu^2}{x}\right)J_\nu(lx) = 0.$$

A multiplication of $J_\nu(lx)$ to the former and $J_\nu(kx)$ to the latter, a subtraction, and an integration over the interval $[0,1]$ yield

$$\int_0^1 xJ_\nu(kx)J_\nu(lx)\,\mathrm{d}x = \frac{1}{k^2 - l^2}[lJ_\nu'(l)J_\nu(k) - kJ_\nu'(k)J_\nu(l)], \quad (2.169)$$

where the prime $(')$ denotes a differentiation with respect to the argument. If k and l are chosen to be two of different zeros of $J_\nu(x)$, *i.e.*, j_m and j_n, $(m \neq n)$ (see §§A.4.2.1c for the details on the zeros of the Bessel function), then the r.h.s. of Eq. (2.169) becomes zero, so that $\{J_\nu(j_n x); (n = 1, 2, 3, \cdots)\}$ constitutes the orthogonal system in the interval $[0,1]$ with the weight function x, *i.e.*,

$$\int_0^1 xJ_\nu(j_m x)J_\nu(j_n x)\,\mathrm{d}x = 0. \quad (2.170)$$

For normalization,

$$\int_0^1 x[J_\nu(j_m x)]^2\mathrm{d}x = \frac{1}{2}[J_{\nu+1}(j_m)]^2 \quad (2.171)$$

is used (the above results may be obtained, *e.g.*, by applying the L'Hospital's theorem to the r.h.s. of Eq. (2.169), which is rewritten by the use of recurrence formulae of the Bessel function). Using this system, we can expand the function $f(x)$ defined in the interval $[0,1]$ as

$$f(x) = \sum_{n=1}^\infty c_n J_\nu(j_n x), \quad (2.172)$$

$$c_m = \frac{2}{[J_{\nu+1}(j_m)]^2}\int_0^1 xf(x)J_\nu(j_m x)\,\mathrm{d}x. \quad (2.173)$$

This expansion fulfills the condition $f = 0$ at $x = 1$ (or at $x = a$ in general by a change of scale), so that it is useful to solve the problems involving the boundary conditions on the circular cylinder. The present expansion is called the **Fourier–Bessel expansion**, which was first introduced by Fourier for $\nu = 0$ case, and later extended to an arbitrary real number $\nu > -1$ by Lommel.

Note that in this expansion (2.172), the following **Parseval's relation** holds:

$$\int_0^1 [f(x)]^2 \, x \, \mathrm{d}x = \frac{1}{2} \sum_{n=1}^{\infty} c_n^2 [J_{\nu+1}(j_n)]^2. \tag{2.174}$$

2.4 Green's Function and the Boundary-value Problems

2.4.1 *Delta function*

The series of functions, as depicted in Fig. 2.9, show the properties in which they have extremely large values at a certain point (or in the neighborhood of that point) whereas they decay to zero with the distance away from that point.

The function attained in the limit $n \to \infty$ is called the **delta function** and is described as $\delta(x)$, which has the following properties:

$$(1) \quad \delta(x) \to \infty \quad \text{at} \quad x = 0, \qquad \delta(x) = 0 \quad \text{at} \quad x \neq 0,$$

$$(2) \quad \int_{-\infty}^{\infty} \delta(x) \, \mathrm{d}x = 1, \tag{2.175}$$

$$(3) \quad \int_{-\infty}^{\infty} f(x)\delta(x - a) \, \mathrm{d}x = f(a).$$

In the third equation, the continuity of $f(x)$ at $x = a$ is assumed. This function is introduced by Dirac, which is a useful tool to describe the ideal properties of the physical substances such as the mass point (a small body that has mass but the size is negligible), point charge (a charged body with a negligible size), and a point force (a force regarded as concentrated to a point), *etc.*[13]

---

[13]Note that the delta function is an even function $\delta(-x) = \delta(x)$, and has the properties such as $\delta(ax) = \delta(x)/a$, $x\,\delta(x) = 0$, and $x\,\delta'(x) = -\delta(x)$, *etc.*

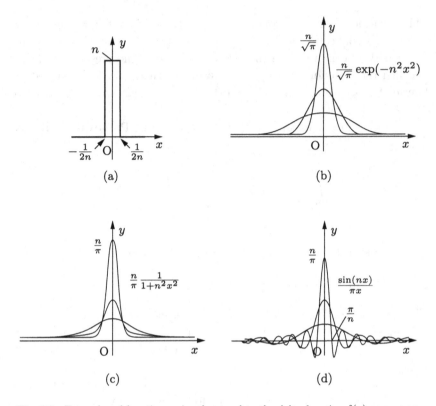

Fig. 2.9   Examples of function series that tend to the delta function $\delta(x)$ as $n \to \infty$.

The mathematical expression $\sin(nx)/\pi x$, shown schematically in Fig. 2.9(d), is rewritten in the limit $n \to \infty$ as

$$\lim_{n\to\infty} \frac{\sin(nx)}{\pi x} = \lim_{n\to\infty} \left(\frac{e^{inx} - e^{-inx}}{2\pi i x} \right)$$

$$= \lim_{n\to\infty} \frac{1}{2\pi} \int_{-n}^{n} e^{ikx} dk = \frac{1}{2\pi} \int_{-\infty}^{\infty} e^{ikx} dk,$$

which is found to be the Fourier transform of the constant $1/\sqrt{2\pi}$ (see §§3.3.2a(iii)-(3)). This relation gives an integral representation of the delta function

$$\delta(x) = \frac{1}{2\pi} \int_{-\infty}^{\infty} e^{ikx} dk. \tag{2.176}$$

If the above equation is integrated with respect to x over the interval

$(-\infty, x]$, we have

$$1(x) \equiv \frac{1}{2\pi} \int_{-\infty}^{\infty} \frac{e^{ikx}}{ik} dk = \begin{cases} 0 & (x < 0) \\ 1 & (x > 0), \end{cases} \tag{2.177}$$

which is called the **unit step function** or the **Heaviside function**.

2.4.2 Adjoint PDE and Green's formulae

A second-order linear PDE is given in general by Eq. (2.9) [or Eq. (2.36)], i.e., $L[u] = G$:

$$L[u] \equiv A\frac{\partial^2 u}{\partial x^2} + 2B\frac{\partial^2 u}{\partial x \partial y} + C\frac{\partial^2 u}{\partial y^2} + D\frac{\partial u}{\partial x} + E\frac{\partial u}{\partial y} + Fu, \tag{2.178}$$

where A, B, $\cdots$, G are functions of x and y. We now define

$$M[v] \equiv \frac{\partial^2(Av)}{\partial x^2} + 2\frac{\partial^2(Bv)}{\partial x \partial y} + \frac{\partial^2(Cv)}{\partial y^2} - \frac{\partial(Dv)}{\partial x} - \frac{\partial(Ev)}{\partial y} + Fv, \tag{2.179}$$

which is called the **adjoint differential operator** of L. We multiply Eq. (2.178) by v, Eq. (2.179) by u, and subtract both sides, which yields

$$vL[u] - uM[v] = P_x + Q_y, \tag{2.180}$$

where the subscript denotes the partial differentiation with respect to this variable unless otherwise stated, and

$$P = A(vu_x - uv_x) + B(vu_y - uv_y) + (D - A_x - B_y)uv,$$

$$Q = B(vu_x - uv_x) + C(vu_y - uv_y) + (E - B_x - C_y)uv.$$

Here, we have taken into account the following calculation, *e.g.*,

$$vAu_{xx} - u(Av)_{xx} = \{(vAu_x)_x - (vA)_x u_x\} - \{[(u(Av_x)]_x - u_x(Av)_x\}$$
$$= [A(vu_x - uv_x)]_x - (A_x uv)_x, \cdots .$$

By integrating the r.h.s. of Eq. (2.180) over the region D within the closed curve C in the xy plane, we have

$$\int\int_D (P_x + Q_y) \, dx \, dy = \int_C (P \, dy - Q \, dx) = \int_C (Pn_x + Qn_y) \, ds, \tag{2.181}$$

where (n_x, n_y) is the direction cosine $n_x = \cos(n, x)$ and $n_y = \sin(n, x)$ of the outward normal vector at the point on C (note here, the subscripts of n_x, n_y refer to the components of the vector, and not refer to the partial differentiations). Consequently, we obtain

$$\int\int_D (vL[u] - uM[v]) \, dx \, dy = \int_C (P \, dy - Q \, dx) = \int_C (Pn_x + Qn_y) \, ds. \tag{2.182}$$

The formula (2.182) is called the **generalized Green's formula**. In particular, if $A = C = 1$, $B = D = E = F = 0$, then

$$L = M = \frac{\partial^2}{\partial x^2} + \frac{\partial^2}{\partial y^2} = \triangle, \quad P = vu_x - uv_x, \quad Q = vu_y - uv_y,$$

so that Eq. (2.182) becomes

$$\int\int_D (v\triangle u - u\triangle v)\,dx\,dy = \int_C \left(v\frac{\partial u}{\partial n} - u\frac{\partial v}{\partial n} \right)\,ds, \qquad (2.183)$$

which is called the **Green's formula**. $\triangleleft$

When $L[u] = M[u]$, $L[u]$ is called the **self-adjoint differential operator**. For this to be realized, we must have

$$M[u] - L[u] = (A_{xx} + 2B_{xy} + C_{yy} - D_x - E_y)u$$
$$+ 2(A_x + B_y - D)u_x + 2(B_x + C_y - E)u_y = 0.$$

Necessary and sufficient conditions for the above requirement are found to be

$$D = A_x + B_y, \qquad E = B_x + C_y.$$

In this case, Eq. (2.178) becomes

$$L[u] = Au_{xx} + 2Bu_{xy} + Cu_{yy} + (A_x + B_y)u_x + (B_x + C_y)u_y + Fu$$
$$= \frac{\partial}{\partial x}\left(A\frac{\partial u}{\partial x} + B\frac{\partial u}{\partial y} \right) + \frac{\partial}{\partial y}\left(B\frac{\partial u}{\partial x} + C\frac{\partial u}{\partial y} \right) + Fu \equiv L^*[u],$$
$$(2.184)$$

where we denote the self-adjoint differential operator $L[u]$ by $L^*[u]$. We also have

$$P = A(vu_x - uv_x) + B(vu_y - uv_y) = v(Au_x + Bu_y) - u(Av_x + Bv_y),$$

$$Q = B(vu_x - uv_x) + C(vu_y - uv_y) = v(Bu_x + Cu_y) - u(Bv_x + Cv_y),$$

so that Eq. (2.182) becomes

$$\int\int_D (vL^*[u] - uL^*[v])\,dx\,dy = \int_C \{[v(Au_x + Bu_y) - u(Av_x + Bv_y)]\,dy$$
$$- [v(Bu_x + Cu_y) - u(Bv_x + Cv_y)]\,dx\}.$$
$$(2.185)$$

Furthermore, if the PDE is elliptic and is given in the standard form, such that $A = C = 1, B = 0$, then $L^*[u]$ is

$$L^*[u] = \frac{\partial^2 u}{\partial x^2} + \frac{\partial^2 u}{\partial y^2} + Fu, \tag{2.186}$$

and Eq. (2.185) becomes

$$\int\int_D (vL^*[u] - uL^*[v])\,\mathrm{d}x\,\mathrm{d}y = \int_C [v(u_x\mathrm{d}y - u_y\mathrm{d}x) - u(v_x\mathrm{d}y - v_y\mathrm{d}x)]$$

$$= \int_C \left(v\frac{\partial u}{\partial n} - u\frac{\partial v}{\partial n} \right)\,\mathrm{d}s. \tag{2.187}$$

◁

Note that surface integrals are reduced to contour integrals in Eq. (2.183) and Eq. (2.187).

2.4.3 Green's function

We shall now introduce the Green's function using a simple example. The ODE that determines the displacement of the string under the distributed load $f(x)$ is described by the following equation (*e.g.* replace the force due to acceleration given by l.h.s. of Eq. (2.1) by the static force due to gravity)

$$\frac{\mathrm{d}^2 u}{\mathrm{d}x^2} = f(x), \tag{2.188}$$

which is a one-dimensional Poisson equation. To obtain the solution, we first consider a string with both ends $x = 0$ and l fixed, to which a concentrated force is applied at a point $P(x = \xi)$ in between. At the point P, the string is displaced by a certain amount, whereas the neighboring parts are displaced in a zigzag manner to connect both ends (see Fig. 2.10). Namely, the displacement of the string is given by

$$u \equiv G(x, \xi) = \begin{cases} x(l - \xi)/l & (0 \le x \le \xi) \\ \xi(l - x)/l & (\xi \le x \le l) \end{cases}. \tag{2.189}$$

As shown in this example, the function G has the following properties:

(1) G satisfies the homogeneous part of the ODE for u at $x \ne \xi$,
    (in the present example, G satisfies $\partial^2 G/\partial x^2 = 0$),
(2) G is continuous at $x = \xi$,
(3) $\partial G/\partial x$ has discontinuity of magnitude -1 at $x = \xi$,
(4) G satisfies the same boundary condition as u
    (in the present example, $G = 0$ at $x = 0,\ l$).

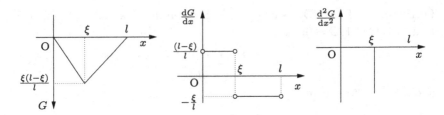

Fig. 2.10   Displacement of the string and the Green's function.

The solution of Eq. (2.188) that satisfies the above conditions is then given by

$$u = -\int_0^l f(\xi)\, G(x, \xi)\, \mathrm{d}\xi. \tag{2.190}$$

where we choose $G(x, \xi)$ for v in Eq. (2.183), and boundary conditions $u = G = 0$ at $x = 0$, l are considered. (Here, surface integral in the l.h.s. and contour integral in the r.h.s. are reduced to line integral and boundary values, respectively, owing to the reduction of dimension.)

Such a function G is called the **Green's function**. Note that the conditions (1) and (3) are included in the following ODE:

$$\frac{\mathrm{d}^2 G}{\mathrm{d}x^2} = -\delta(x - \xi), \tag{2.191}$$

where $\delta(x)$ is the delta function mentioned before (§§2.4.1).   ◁

In general, to obtain the solution of the linear inhomogeneous ODE:

$$L[u] = f(x), \tag{2.192}$$

we first seek a solution $G(x, \xi)$ of the equation

$$L[G(x, \xi)] = -\delta(x - \xi), \tag{2.193}$$

that satisfies the same boundary condition as u. The latter is the Green's function. Then, the solution of Eq. (2.192) is given by

$$u = -\int f(\xi)\, G(x, \xi)\, \mathrm{d}\xi. \tag{2.194}$$

2.4.4   *Principal solution of the Laplace equation*

We consider **principal solutions** of the Laplace equation. The principal solution implies the one that satisfies the discontinuous condition at a singular point.

a. One-dimensional Laplace equation

The solution of

$$\frac{\mathrm{d}^2 G}{\mathrm{d}x^2} = -\delta(x - \xi), \tag{2.195}$$

that satisfies the boundary condition was already given in §§2.4.3. However, we shall focus here on finding the principal solution, and compare it with that already obtained. We first note that the r.h.s. of Eq. (2.195) is given by (see Eq. (2.176))

$$-\delta(x - \xi) = -\frac{1}{2\pi} \int_{-\infty}^{\infty} e^{ik(x-\xi)} \mathrm{d}k,$$

which reminds us of the principal solution G of the form

$$G(x, \xi) = \frac{1}{2\pi} \int_{-\infty}^{\infty} g(k) e^{ik(x-\xi)} \mathrm{d}k. \tag{2.196}$$

If we substitute the latter for Eq. (2.195), and compare the like terms, we obtain

$$g(k) = \frac{1}{k^2}. \tag{2.197}$$

Hence, we obtain

$$G(x, \xi) = \frac{1}{2\pi} \int_{-\infty}^{\infty} \frac{e^{ik(x-\xi)}}{k^2} \mathrm{d}k. \tag{2.198}$$

For further calculation, we make use of the unit step function (2.177). Namely, the integration of the latter with respect to x over the interval from $-\infty$ to x:

$$\int_{-\infty}^{x} 1(x - \xi) \, \mathrm{d}x = \int_{-\infty}^{x} \frac{1}{2\pi} \int_{-\infty}^{\infty} \frac{e^{ik(x-\xi)}}{ik} \, \mathrm{d}k \, \mathrm{d}x$$

$$= -\frac{1}{2\pi} \int_{-\infty}^{\infty} \frac{e^{ik(x-\xi)}}{k^2} \, \mathrm{d}k = -G,$$

yields

$$G(x, \xi) = -\int_{-\infty}^{x} 1(x - \xi) \, \mathrm{d}x = \begin{cases} 0 & (x \leq \xi) \\ \xi - x & (x \geq \xi) \end{cases}. \tag{2.199}$$

Compared with Eq. (2.189), the results (2.199) have a difference of $x(l-\xi)/l$, which is a solution of the homogeneous part of Eq. (2.195). As revealed here, the principal solution, in general, has the indeterminacy of the homogeneous solution of the relevant ODE.

b. Two-dimensional Laplace equation

The principal solution of the two-dimensional Laplace equation G satisfies the following equation:

$$\Delta G(x, y; \xi, \eta) = -\delta(x - \xi)\,\delta(y - \eta). \tag{2.200}$$

Note that the r.h.s. of the above equation is given by

$$-\frac{1}{(2\pi)^2} \int_{-\infty}^{\infty} \int_{-\infty}^{\infty} e^{i[k_x(x-\xi)+k_y(y-\eta)]} dk_x\, dk_y,$$

so that we expect the principal solution G to be of the form

$$G(x, y; \xi, \eta) = \frac{1}{(2\pi)^2} \int_{-\infty}^{\infty} \int_{-\infty}^{\infty} g(k_x, k_y)\, e^{i[k_x(x-\xi)+k_y(y-\eta)]} dk_x\, dk_y. \tag{2.201}$$

After a substitution of the latter for Eq. (2.200), and a comparison of like terms, we obtain

$$g(k_x, k_y) = \frac{1}{k_x^2 + k_y^2}. \tag{2.202}$$

Hence, we have

$$G = \frac{1}{(2\pi)^2} \int_{-\infty}^{\infty} \int_{-\infty}^{\infty} \frac{e^{i(k_x u + k_y v)}}{k_x^2 + k_y^2} dk_x\, dk_y, \tag{2.203}$$

where we put $u = x - \xi$ and $v = y - \eta$ for brevity.

We now calculate Eq. (2.203) by the use of a contour integral in the complex plane. (For complex integrals, see *e.g.*, [Brown and Churchill (2009); Fujiwara (2013)], *etc.*)

We first calculate the integral with respect to k_x:

$$I = \frac{1}{2\pi} \int_{-\infty}^{\infty} \frac{e^{ik_x u}}{k_x^2 + k_y^2} dk_x.$$

To do this, we extend k_x to the complex variable, and consider a contour integral

$$J = \frac{1}{2\pi} \int_C \frac{e^{ik_x u}}{k_x^2 + k_y^2} dk_x,$$

where we take a closed contour C as shown in Fig. 2.11(a). The latter contour consists of the line segment AB on the real axis from $k_x = -R$ to $k_x = R$, and the semi-circle C_R of radius R in the upper-half plane to complete a closed path. The singularity inside the closed path C is a simple pole at $k_x = i|k_y| \equiv i\alpha$ ($\alpha > 0$), so that we have

$$J = \frac{1}{2\pi} 2\pi i \operatorname{Res}(i\alpha) = \frac{1}{2\alpha} e^{-\alpha u}$$

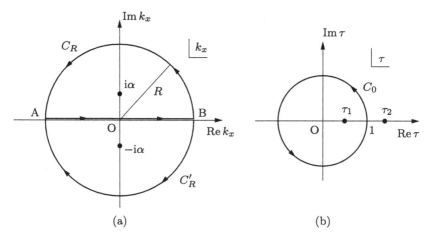

Fig. 2.11 Integral path for calculating the Green's function of the two- and three-dimensional Laplace equation.

by the residue theorem. Meanwhile, the contour integral along the semicircle C_R tends to zero as $R \to \infty$ for $u > 0$ by the Jordan's lemma. Therefore, we have

$$J = \frac{1}{2\pi} \int_C \cdots = \frac{1}{2\pi} \left(\int_{AB} \frac{e^{ik_x u}}{k_x^2 + k_y^2} \, dk_x + \int_{C_R} \cdots \right) \to I \quad (R \to \infty),$$

and hence $I = e^{-\alpha u}/(2\alpha)$. If $u < 0$, we perform the complex integral J along the closed contour C, which consists of the line segment AB and the semi-circle C_R' of radius R in the lower-half plane. In this case, the contribution of the integral along the path C_R' tends to zero as $R \to \infty$, so that $J \to I$ $(R \to \infty)$. Since the singularity inside the contour C is a simple pole at $k_x = -i\alpha$, the residue theorem gives

$$J = -2\pi i \frac{1}{2\pi} \mathrm{Res}(-i\alpha) = \frac{1}{2\alpha} e^{\alpha u} = I$$

(note that the sense of rotation around the singular point is clockwise in the latter case). Accordingly, we obtain the integral I, irrespective of the sign of u:

$$I = \frac{1}{2\pi} \int_{-\infty}^{\infty} \frac{e^{ik_x u}}{k_x^2 + k_y^2} \, dk_x = \frac{1}{2|k_y|} e^{-|k_y u|}. \tag{2.204}$$

Using the above result, we have

$$\begin{aligned} G &= \frac{1}{2\pi} \int_{-\infty}^{\infty} \frac{1}{2|k_y|} e^{-|k_y u| + ik_y v} \, dk_y \\ &= \frac{1}{2\pi} \int_0^{\infty} \frac{1}{k_y} e^{-k_y |u|} \cos(k_y v) \, dk_y. \end{aligned} \tag{2.205}$$

We further calculate the integral with respect to k_y,[14] which yields

$$G(x, y; \xi, \eta) = -\frac{1}{4\pi} \log(u^2 + v^2) = -\frac{1}{2\pi} \log r, \qquad (2.206)$$

where we put $r = \sqrt{u^2 + v^2} = \sqrt{(x - \xi)^2 + (y - \eta)^2}$.

c. Three-dimensional Laplace equation

The principal solution G of the three-dimensional Laplace equation satisfies the following equation:

$$\triangle G(x, y, z; \xi, \eta, \zeta) = -\delta(x - \xi)\,\delta(y - \eta)\,\delta(z - \zeta). \qquad (2.207)$$

Similar to one- and two-dimensional cases, we remark that the r.h.s. is given by

$$-\frac{1}{(2\pi)^3} \int_{-\infty}^{\infty} \int_{-\infty}^{\infty} \int_{-\infty}^{\infty} e^{i[k_x(x-\xi)+k_y(y-\eta)+k_z(z-\zeta)]}\,dk_x dk_y dk_z,$$

so that we assume G of the form

$$G(x, y, z, t; \xi, \eta, \zeta, \tau) = \frac{1}{(2\pi)^3} \int_{-\infty}^{\infty} \int_{-\infty}^{\infty} \int_{-\infty}^{\infty} g(k_x, k_y, k_z)$$
$$\times e^{i[k_x(x-\xi)+k_y(y-\eta)+k_z(z-\zeta)]}\,dk_x\,dk_y\,dk_z. \qquad (2.208)$$

A substitution of the latter for Eq. (2.207) and a comparison of like terms yield

$$g(k_x, k_y, k_z) = \frac{1}{k_x{}^2 + k_y{}^2 + k_z{}^2}. \qquad (2.209)$$

Hence, we have

$$G = \frac{1}{(2\pi)^3} \int_{-\infty}^{\infty} \int_{-\infty}^{\infty} \int_{-\infty}^{\infty} \frac{e^{i(k_x u+k_y v+k_z w)}}{k_x{}^2 + k_y{}^2 + k_z{}^2}\,dk_x\,dk_y\,dk_z, \qquad (2.210)$$

where we put $u = x - \xi$, $v = y - \eta$, $w = z - \zeta$ for brevity.

---

[14]The calculation of

$$I = \int_0^{\infty} \frac{1}{x}e^{-\beta x}\cos(\alpha x)\,dx \qquad (\beta > 0)$$

may be performed, *e.g.*, with the use of repeated integration by parts.

$$J \equiv \frac{dI}{d\alpha} = -\int_0^{\infty} \sin(\alpha x)\,e^{-\beta x}\,dx = \left[\frac{\cos(\alpha x)}{\alpha}e^{-\beta x}\right]_0^{\infty} + \frac{\beta}{\alpha}\int_0^{\infty} \cos(\alpha x)\,e^{-\beta x}\,dx$$
$$= -\frac{1}{\alpha} + \frac{\beta}{\alpha}\int_0^{\infty} \cos(\alpha x)\,e^{-\beta x}\,dx = \dots = -\frac{1}{\alpha} - \frac{\beta^2}{\alpha^2}J,$$

from which we have

$$J = -\frac{\alpha}{\alpha^2 + \beta^2} \quad \rightarrow \quad I = -\frac{1}{2}\log(\alpha^2 + \beta^2) + \text{constant}.$$

The constant of integration may be neglected for obtaining the Green's function.

Although the integral of Eq. (2.210) may be performed by using the spherical coordinate system, we do it here by means of the contour integral, as has been dealt with in the two-dimensional case. We first calculate the integral with respect to k_x

$$I = \int_{-\infty}^{\infty} \frac{e^{ik_x u}}{k_x{}^2 + \alpha^2}\, dk_x,$$

where $\alpha = \sqrt{k_y{}^2 + k_z{}^2}$. The calculation is almost the same as we have shown in the two-dimensional case (§§2.4.4b), so that the integral I is, irrespective of the sign of u,

$$I = \int_{-\infty}^{\infty} \frac{e^{ik_x u}}{k_x{}^2 + \alpha^2}\, dk_x = \frac{\pi}{\alpha} e^{-\alpha|u|}. \tag{2.211}$$

Next, we substitute Eq. (2.211) for Eq. (2.210):

$$G = \frac{1}{8\pi^2} \int_{-\infty}^{\infty} \int_{-\infty}^{\infty} \frac{1}{\sqrt{k_y{}^2 + k_z{}^2}}$$

$$\times \exp\left(i(k_y v + k_z w) - |u|\sqrt{k_y{}^2 + k_z{}^2} \right) dk_y dk_z.$$

By putting $k_y = k\cos\theta$, $k_z = k\sin\theta$, $k^2 = k_y{}^2 + k_z{}^2$, we have

$$G = \frac{1}{8\pi^2} \int_0^{\infty} \int_0^{2\pi} \exp\left[-k(\,|u| - iv\cos\theta - iw\sin\theta)\right]\, dk d\theta$$

$$= \frac{1}{8\pi^2} \int_0^{2\pi} \frac{1}{|u| - iv\cos\theta - iw\sin\theta}\, d\theta \tag{2.212}$$

$$= \frac{1}{8\pi^2} \int_0^{2\pi} \frac{1}{|u| - i\beta\sin(\theta + \gamma)}\, d\theta,$$

where $\beta^2 = v^2 + w^2$ and $\gamma = \arctan(v/w)$. To calculate Eq. (2.212), we furthermore put $e^{i(\theta+\gamma)} = \tau$. Then we have

$$d\theta = \frac{d\tau}{i\tau}, \qquad \sin(\theta + \gamma) = \frac{1}{2i}\left(\tau - \frac{1}{\tau}\right),$$

where the integral over θ is realized by the contour integral on the unit circle C_0 in the complex τ plane (see Fig. 2.11(b)). By this transformation, we have

$$G = \frac{1}{8\pi^2} \int_{C_0} \frac{1}{|u| - \dfrac{\beta}{2}\left(\tau - \dfrac{1}{\tau}\right)}\, \frac{d\tau}{i\tau} = -\frac{1}{4\pi^2 i\beta} \int_{C_0} \frac{d\tau}{\tau^2 - \dfrac{2|u|}{\beta}\tau - 1}.$$

Here, $\tau^2 - (2|u|/\beta)\tau - 1 = 0$ has two different real solutions

$$\tau = \frac{|u|}{\beta} \mp \sqrt{\frac{u^2}{\beta^2} + 1} = \begin{cases} \tau_1 \\ \tau_2 \end{cases},$$

where only τ_1 exists inside C_0. By the residue theorem, we can calculate the integral

$$G = -\frac{1}{4\pi^2 i\beta} 2\pi i \operatorname{Res}(\tau_1) = \frac{1}{2\pi\beta(\tau_2 - \tau_1)}$$

$$= \frac{1}{4\pi\sqrt{u^2 + \beta^2}} = \frac{1}{4\pi\sqrt{u^2 + v^2 + w^2}},$$

from which we obtain the principal solution

$$G = \frac{1}{4\pi R}, \quad R = \sqrt{(x - \xi)^2 + (y - \eta)^2 + (z - \zeta)^2}. \tag{2.213}$$

$\triangleleft$

From the view point of physics, Eq. (2.213) has the same form as the electrostatic potential owing to a point charge, or as the potential created by a mass point under the law of universal gravitation. The potential owing to the distributed electrostatic charge or distributed mass $q(\boldsymbol{x})$ is then given by

$$u(\boldsymbol{x}) = -\int_V \frac{q(\boldsymbol{x}')}{4\pi|\boldsymbol{x} - \boldsymbol{x}'|} \, d\boldsymbol{x}', \tag{2.214}$$

where q includes the proportional constant that originates in respective physical quantities and the choice of unit systems. Similarly, Eq. (2.206) in the two-dimensional case corresponds to, *e.g.*, electrostatic potential due to the line source of a charge in electrostatics.

Owing to the linearity of the Laplace equation, the functions in which the solution of the latter are differentiated many times with respect to x, y, z are also the solution of the Laplace equation. For this reason, the fundamental solution with a simpler configuration or higher symmetry has greater importance. The principal solutions of the Laplace equation in two and three dimensions are also obtained by using the cylindrical or spherical coordinate system, respectively, in which the Green's theorem or the Gauss' theorem is used.[15]

---

[15]Consider the two-dimensional circular symmetric Laplace equation in a polar coordinate system (r, θ) with its origin at $r \equiv \boldsymbol{x} - \boldsymbol{x}' = 0$. The equation is given

2.4.5 Adjoint Green's function and reciprocity principle

As stated in §§2.4.2, a second-order linear PDE is in general described in the form (2.9) [or Eq. (2.36)]. We denoted the partial differential operator of this equation by $L[u]$ (see Eq. (2.178)), and the adjoint partial differential operator by $M[u]$ (see Eq. (2.179)). We now choose u and v as

$$u = G(x, y; \xi, \eta), \qquad v = H(x, y; \xi, \eta), \tag{2.219}$$

where

$$L[G(x, y; \xi, \eta)] = -\delta(x - \xi)\,\delta(y - \eta), \tag{2.220}$$

and

$$M[H(x, y; \xi, \eta)] = -\delta(x - \xi)\,\delta(y - \eta). \tag{2.221}$$

---

by

$$\frac{1}{r}\frac{d}{dr}\left(r\frac{dG}{dr}\right) = -\delta(\boldsymbol{r}). \tag{2.215}$$

We first obtain the solution $G = C\log r$ (where C is a constant and $r = |\boldsymbol{r}|$), which is valid for $r \neq 0$. We then apply the Green's formula (2.183) with $u = -1$, $v = G$, in which integration over the circular region of radius ϵ centered at the origin is considered. Then, we have

$$\int_S \triangle G \, dS = \int_C \frac{\partial G}{\partial r}\, ds = \int_C \frac{C}{r}\, ds = 2\pi C = -1,$$

from which we have $C = -1/(2\pi)$, and hence

$$G(\boldsymbol{x}, \boldsymbol{x}') = -\frac{1}{2\pi}\log r = \frac{1}{2\pi}\log\frac{1}{|\boldsymbol{x} - \boldsymbol{x}'|}. \tag{2.216}$$

Similarly, consider the spherically symmetric Laplace equation in a spherical coordinate system with its origin at $\boldsymbol{r} \equiv \boldsymbol{x} - \boldsymbol{x}' = \boldsymbol{0}$. The equation is given by

$$\triangle G(\boldsymbol{x}, \boldsymbol{x}') \equiv \frac{1}{r^2}\frac{d}{dr}\left(r^2\frac{dG}{dr}\right) = -\delta(\boldsymbol{r}). \tag{2.217}$$

We first solve the equation $\triangle G = 0$ that is valid at $r \equiv |\boldsymbol{x} - \boldsymbol{x}'| \neq 0$, which yields $G = C/r$ (where C is a constant). Next, we apply the Gauss' theorem (three-dimensional version of the Green's theorem), in which the integration over the sphere of radius ϵ centered at the origin is performed. Then we have,

$$\int_V \triangle G \, dV = \int_V \nabla \cdot \nabla G \, dV = \int_S \frac{\partial G}{\partial r}\, dS = -\int_S \frac{C}{r^2}\, dS = -\frac{C}{\epsilon^2}4\pi\epsilon^2 = -4\pi C.$$

Since the integral of the r.h.s. of Eq. (2.217) gives -1, we have $C = 1/4\pi$, and hence

$$G(\boldsymbol{x}, \boldsymbol{x}') = \frac{1}{4\pi r} = \frac{1}{4\pi|\boldsymbol{x} - \boldsymbol{x}'|}. \tag{2.218}$$

If Eq. (2.182) is zero, H is called the **adjoint Green's function**. In this case, by putting $u = G(x, y; \xi_1, \eta_1)$ and $v = H(x, y; \xi_2, \eta_2)$, we have the following relation:

$$\int\int_D (H(x, y; \xi_2, \eta_2)\, L[G(x, y; \xi_1, \eta_1)]$$

$$-G(x, y; \xi_1, \eta_1)\, M[H(x, y; \xi_2, \eta_2)])\, dx\, dy = 0,$$

i.e.,

$$\int\int_D [H(x, y; \xi_2, \eta_2)\, \delta(x - \xi_1)\, \delta(y - \eta_1)$$

$$-G(x, y; \xi_1, \eta_1)\, \delta(x - \xi_2)\, \delta(y - \eta_2)]\, dx\, dy = 0.$$

If (ξ_1, η_1) and (ξ_2, η_2) are inner points of the region D, we have the relation

$$H(\xi_1, \eta_1; \xi_2, \eta_2) = G(\xi_2, \eta_2; \xi_1, \eta_1). \tag{2.222}$$

In other words, if the variables (ξ_1, η_1) and the variables (ξ_2, η_2) are exchanged, the Green's function becomes its adjoint Green's function. In particular, if they are self-adjoint to each other, so that $G = H$, we have the relation

$$G(\xi_1, \eta_1; \xi_2, \eta_2) = G(\xi_2, \eta_2; \xi_1, \eta_1), \tag{2.223}$$

which is called the **reciprocity principle** of the Green's function. The Green's functions thus far dealt with, *e.g.*, (2.189), (2.206), and (2.213), satisfy this relation. In contrast, the Green's function for diffusion equation Eq. (2.239), and hence (2.244), (2.249), etc. that are dealt with later are not self-adjoint.

2.4.6   *Principal solution of the wave equation*

a. One-dimensional wave equation

The principal solution satisfies the following equation:

$$L[G(x, t; \xi, \tau)] \equiv \frac{\partial^2 G}{\partial x^2} - \frac{1}{c^2}\frac{\partial^2 G}{\partial t^2} = -\delta(x - \xi)\, \delta(t - \tau). \tag{2.224}$$

Taking into account that the r.h.s. is expressed by

$$\delta(x - \xi) = \frac{1}{2\pi}\int_{-\infty}^{\infty} e^{ik(x-\xi)}dk, \qquad \delta(t - \tau) = \frac{1}{2\pi}\int_{-\infty}^{\infty} e^{-i\omega(t-\tau)}d\omega,$$

we assume that the principal solution G is given by

$$G(x, t; \xi, \tau) = \frac{1}{(2\pi)^2}\int_{-\infty}^{\infty}\int_{-\infty}^{\infty} g(k, \omega)\, e^{i[k(x-\xi)-\omega(t-\tau)]}dk\, d\omega. \tag{2.225}$$

Substitution of the latter for Eq. (2.224) and a comparison of like terms yield

$$g(k,\omega) = \frac{c^2}{k^2c^2 - \omega^2}, \qquad (2.226)$$

so that the solution is given by

$$G(x,t;\xi,\tau) = \frac{1}{(2\pi)^2} \int_{-\infty}^{\infty} \int_{-\infty}^{\infty} \frac{c^2}{k^2c^2 - \omega^2}\, e^{i[k(x-\xi)-\omega(t-\tau)]} dk\, d\omega. \quad (2.227)$$

We now perform the integral with respect to ω:

$$I \equiv \frac{c^2}{2\pi} \int_{-\infty}^{\infty} \frac{e^{-i\omega(t-\tau)}}{\omega_0{}^2 - \omega^2}\, d\omega = \frac{c^2}{4\pi\omega_0} \int_{-\infty}^{\infty} \left(\frac{e^{-i\omega(t-\tau)}}{\omega + \omega_0} - \frac{e^{-i\omega(t-\tau)}}{\omega - \omega_0} \right) d\omega,$$
$$(2.228)$$

where we put $\omega_0 = kc$. We extend ω to a complex variable, and perform the contour integral along a closed path consisting of the real axis and the semi-circle of large radius in the upper- or lower-half planes (see Fig. 2.12(a)).

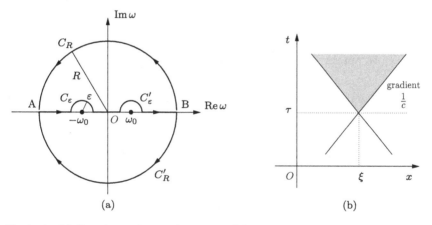

(a)   (b)

Fig. 2.12   (a) Causality and integral path, and (b) region of influence in wave equations.

The choice of the path needs some consideration, because of the presence of simple poles at $\omega = \pm\,\omega_0$ on the real axis. We should also pay attention to the findings that when the wave is created at $t = \tau$, "nothing is recognized (*i.e.*, $I = 0$)" for $t < \tau$. The latter is a requirement of the **causality** (see *e.g.*, [Byron and Fuller (1969)] in more detail). To fulfill the causality requirement, we modify the integral path such that the singularities $\omega = \pm\,\omega_0$ are avoided by small semi-circles C_ϵ and C'_ϵ on the side of the upper-half plane. Then, no singularities are present inside the closed path consisting of the real axis, C_R, C_ϵ and C'_ϵ, and hence the causality requirement for $t < \tau$ is satisfied. (Note that the contribution of the integral on C_R tends to zero as $R \to \infty$.) For $t > \tau$, the integral on C'_R tends to zero as $R \to \infty$, whereas the contribution from the simple poles $\omega = \pm \omega_0$ inside the closed path consisting of the real axis, $C_\epsilon, C'_\epsilon, C'_R$ remains. The latter is calculated by the residue theorem, which yields

$$I = \frac{c^2}{\omega_0} \sin[\omega_0(t - \tau)] = \frac{c}{k} \sin[kc(t - \tau)]. \tag{2.229}$$

Consequently, we obtain

$$G(x, t; \xi, \tau) = \begin{cases} 0 & (t < \tau) \\ \dfrac{c}{2\pi} \displaystyle\int_{-\infty}^{\infty} \dfrac{1}{k} \sin[kc(t - \tau)] \, e^{ik(x - \xi)} \, dk & (t > \tau) \end{cases}. \tag{2.230}$$

In terms of the unit step function (see Eq. (2.177)), the expression for $t > \tau$ is rewritten as

$$G(x, t; \xi, \tau) = \frac{c}{2} \left[\mathbf{1}\big(x - \xi + c(t - \tau)\big) - \mathbf{1}\big(x - \xi - c(t - \tau)\big) \right]$$
$$= \frac{c}{2} \mathbf{1}\big(c(t - \tau) - |x - \xi|\big),$$

so that, we obtain

$$G(x, t; \xi, \tau) = \begin{cases} \dfrac{c}{2} & \text{for} \quad t > \tau \quad \text{and} \quad |x - \xi| < c(t - \tau) \\ 0 & \text{otherwise} \end{cases}. \tag{2.231}$$

From a view point of physics, the wave created at $t = \tau$, $x = \xi$ propagates to the region as shown by the shaded area of Fig. 2.12(b), which is the region of influence of the wave (or event) given by Eq. (2.231).

b. Three-dimensional wave equation

The principal solution G of the three-dimensional wave equation satisfies

$$\triangle G - \frac{1}{c^2} \frac{\partial^2 G}{\partial t^2} = -\delta(\boldsymbol{x} - \boldsymbol{\xi}) \, \delta(t - \tau), \tag{2.232}$$

where $\boldsymbol{x} = (x, y, z)$ and $\boldsymbol{\xi} = (\xi, \eta, \zeta)$. Taking into account that the r.h.s. is expressed by

$$\delta(\boldsymbol{x} - \boldsymbol{\xi}) = \frac{1}{(2\pi)^3} \int_{-\infty}^{\infty} e^{i\boldsymbol{k} \cdot (\boldsymbol{x} - \boldsymbol{\xi})} d\boldsymbol{k}, \qquad \delta(t - \tau) = \frac{1}{2\pi} \int_{-\infty}^{\infty} e^{-i\omega(t - \tau)} d\omega,$$

we assume the principal solution G of the form

$$G(\boldsymbol{x}, t; \boldsymbol{\xi}, \tau) = \frac{1}{(2\pi)^4} \int_{-\infty}^{\infty} \int_{-\infty}^{\infty} g(\boldsymbol{k}, \omega) \, e^{i(\boldsymbol{k} \cdot \boldsymbol{R} - \omega T)} d\boldsymbol{k} \, d\omega, \tag{2.233}$$

where $R = x - \xi$ and $T = t - \tau$. Substitution of the latter for Eq. (2.232) and a comparison of like terms yield

$$g(\boldsymbol{k}, \omega) = \frac{c^2}{k^2 c^2 - \omega^2}, \qquad k = |\boldsymbol{k}|, \qquad (2.234)$$

so that the solution is given by

$$G(\boldsymbol{x}, t; \boldsymbol{\xi}, \tau) = \frac{1}{(2\pi)^4} \int_{-\infty}^{\infty} \int_{-\infty}^{\infty} \frac{c^2}{k^2 c^2 - \omega^2} e^{i(\boldsymbol{k} \cdot \boldsymbol{R} - \omega T)} d\boldsymbol{k} \, d\omega. \qquad (2.235)$$

We shall perform the above integral. Note that the integral with respect to ω:

$$I \equiv \frac{c^2}{2\pi} \int_{-\infty}^{\infty} \frac{e^{-i\omega T}}{k^2 c^2 - \omega^2} d\omega,$$

is the same as Eq. (2.228), so that we have $I = 0$ for $T < 0$, and $I = (c/k)\sin(kcT)$ for $T > 0$. The substitution of I for Eq. (2.235) yields

$$G = \frac{1}{(2\pi)^3} \int_{-\infty}^{\infty} \frac{c}{k} \sin(kcT) \, e^{i\boldsymbol{k} \cdot \boldsymbol{R}} d\boldsymbol{k}. \qquad (2.236)$$

Here, the integral with respect to $\boldsymbol{k}$ is performed over the entire $\boldsymbol{k}$ space, so that we make use of the spherical coordinate system in $\boldsymbol{k}$ space. We choose $\boldsymbol{R}$ as a basic axis, and denote the angle between $\boldsymbol{k}$ and $\boldsymbol{R}$ by θ, and the angle around this axis by φ. Then we have $\boldsymbol{k} \cdot \boldsymbol{R} = kR\cos\theta$, $d\boldsymbol{k} = k^2 \sin\theta \, dk \, d\theta \, d\varphi$, where $R = |\boldsymbol{R}| = \sqrt{(x - \xi)^2 + (y - \eta)^2 + (z - \zeta)^2}$, $|\boldsymbol{k}| = k$, and

$$G = \frac{1}{(2\pi)^3} \int_0^{2\pi} d\varphi \int_0^{\pi} \sin\theta \, d\theta \int_0^{\infty} k^2 \left(\frac{c}{k} \sin(kcT) \right) e^{ikR\cos\theta} dk. \qquad (2.237)$$

Evidently, the integral with respect to φ is given by multiplication of 2π. To perform the integral with respect to θ, we put $\cos\theta = x$, which yields

$$\int_0^{\pi} \sin\theta \, e^{ikR\cos\theta} \, d\theta = \int_{-1}^{1} e^{ikRx} \, dx = \frac{2}{kR} \sin(kR),$$

so that we have

$$\begin{aligned}
G &= \frac{c}{4\pi^2 R} \int_0^{\infty} 2\sin(kR)\sin(kcT) \, dk \\
&= \frac{c}{4\pi^2 R} \int_{-\infty}^{\infty} \frac{(e^{ikR} - e^{-ikR})}{2i} \frac{(e^{ikcT} - e^{-ikcT})}{2i} \, dk \\
&= -\frac{c}{8\pi R} \frac{1}{2\pi} \int_{-\infty}^{\infty} \left(e^{ik(R+cT)} - e^{ik(R-cT)} - e^{-ik(R-cT)} + e^{-ik(R+cT)} \right) dk \\
&= -\frac{c}{4\pi R} \left[\delta(R + cT) - \delta(R - cT) \right] = \frac{c}{4\pi R} \delta(R - cT).
\end{aligned}$$

In the last line of the equation, we have taken into account of the parity $\delta(-x) = \delta(x)$, and the property that $\delta(R + cT) = 0$ because we consider the case $R > 0$, $T > 0$. In this way, we obtain the principal solution

$$G(x, t; \xi, \tau) = \begin{cases} 0 & (t < \tau) \\ \dfrac{c}{4\pi R} \, \delta\big(R - c(t - \tau)\big) & (t > \tau) \end{cases}. \tag{2.238}$$

The solution shows that the wave initiated at $t = \tau$ at the point (ξ, η, ζ) propagates to the spherical surface $R = c(t - \tau)$ at time t, with its magnitude decreasing as $1/R$.

2.4.7 *Principal solution of the diffusion equation*

a. One-dimensional diffusion equation

The principal solution of the one-dimensional diffusion equation satisfies the following equation

$$L[G(x, t; \xi, \tau)] \equiv D \frac{\partial^2 G}{\partial x^2} - \frac{\partial G}{\partial t} = -\delta(x - \xi)\,\delta(t - \tau), \tag{2.239}$$

where $D > 0$. Taking into account that the r.h.s. is expressed by

$$-\frac{1}{(2\pi)^2} \int_{-\infty}^{\infty} \int_{-\infty}^{\infty} e^{i[k(x-\xi) - \omega(t-\tau)]} dk \, d\omega,$$

we assume G of the form

$$G(x, t; \xi, \tau) = \frac{1}{(2\pi)^2} \int_{-\infty}^{\infty} \int_{-\infty}^{\infty} g(k, \omega) \, e^{i[k(x-\xi) - \omega(t-\tau)]} dk \, d\omega. \tag{2.240}$$

Substitution of the latter for Eq. (2.239) and a comparison of like terms yield

$$g(k, \omega) = \frac{-1}{i\omega - Dk^2}, \tag{2.241}$$

so that the solution is given by

$$G(x, t; \xi, \tau) = \frac{1}{(2\pi)^2} \int_{-\infty}^{\infty} \int_{-\infty}^{\infty} \frac{-1}{i\omega - Dk^2} \, e^{i[k(x-\xi) - \omega(t-\tau)]} dk \, d\omega. \tag{2.242}$$

We now perform the integral with respect to ω:

$$I = \frac{1}{2\pi} \int_{-\infty}^{\infty} \frac{-1}{i\omega - Dk^2} \, e^{-i\omega(t-\tau)} d\omega. \tag{2.243}$$

To do so, we extend ω to a complex variable, and consider the contour integral along a closed path that consists of the real axis and a semi-circle

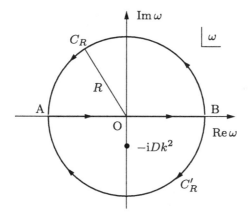

Fig. 2.13 Integral path and causality for the diffusion equation.

in the upper- or lower-half planes (see Fig. 2.13). A simple pole exists at $\omega = -iDk^2$.

We first consider the case $t < \tau$, and perform the contour integral along the path connecting the real axis and the semi-circle C_R in the upper-half plane. For $t < \tau$, the contribution from the path C_R tends to zero as $R \to \infty$, whereas there is no singularity inside the contour, so that the integral I is zero. In the case $t > \tau$, we perform the contour integral along the path that connects the real axis and the semi-circle C_R' in the lower-half plane. Here, the integral over C_R' tends to zero as $R \to \infty$, whereas a simple pole at $\omega = -iDk^2$ gives a residue $e^{-D(t-\tau)k^2}$. Consequently, we have $G = 0$ for $t < \tau$, and

$$
\begin{aligned}
G(x,t;\xi,\tau) &= \frac{1}{2\pi} \int_{-\infty}^{\infty} e^{-D(t-\tau)k^2 + ik(x-\xi)} dk \\
&= \frac{1}{\sqrt{4\pi D(t-\tau)}} \exp\left(-\frac{(x-\xi)^2}{4D(t-\tau)}\right),
\end{aligned}
\tag{2.244}
$$

for $t > \tau$.[16] The space-time behavior of $G(x,t;0,0)$ for $t > 0$ is shown in Fig. 2.14.

---

[16]For the integral with real constant $a \,(> 0)$ and b:

$$
\int_{-\infty}^{\infty} e^{-ax^2 + ibx} dx = \sqrt{\frac{\pi}{a}} \exp\left(-\frac{b^2}{4a}\right),
$$

see, *e.g.*, [Brown and Churchill (2009); Fujiwara (2013)].

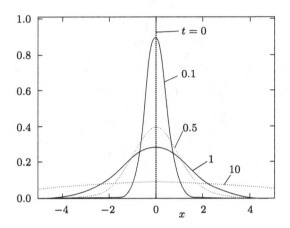

Fig. 2.14 Space-time development by diffusion ($D = 1$, $\tau = 0$, $\xi = 0$).

b. Three-dimensional diffusion equation

The principal solution of the three-dimensional diffusion equation is similarly obtained. The basic equation is

$$D \triangle G - \frac{\partial G}{\partial t} = -\delta(\boldsymbol{x} - \boldsymbol{\xi})\,\delta(t - \tau), \qquad (2.245)$$

where $D > 0$. For brevity, we denote $\boldsymbol{R} = \boldsymbol{x} - \boldsymbol{\xi}$, $\boldsymbol{x} = (x, y, z)$, $\boldsymbol{\xi} = (\xi, \eta, \zeta)$ and $T = t - \tau$. By extending the one-dimensional case, we have

$$\delta(\boldsymbol{R})\,\delta(T) = \frac{1}{(2\pi)^4} \int_{-\infty}^{\infty} \int_{-\infty}^{\infty} e^{i(\boldsymbol{k}\cdot\boldsymbol{R} - \omega T)}\,d\boldsymbol{k}\,d\omega,$$

and

$$G(\boldsymbol{x}, t; \boldsymbol{\xi}, \tau) = \frac{1}{(2\pi)^4} \int_{-\infty}^{\infty} \int_{-\infty}^{\infty} g(\boldsymbol{k}, \omega)\,e^{i(\boldsymbol{k}\cdot\boldsymbol{R} - \omega T)}\,d\boldsymbol{k}\,d\omega, \qquad (2.246)$$

which are substituted for Eq. (2.245) to obtain g:

$$g(\boldsymbol{k}, \omega) = \frac{-1}{i\omega - Dk^2}, \qquad (2.247)$$

the latter being the same as Eq. (2.241) except for $k = |\boldsymbol{k}|$ here. Then the solution is

$$G(\boldsymbol{x}, t; \boldsymbol{\xi}, \tau) = \frac{1}{(2\pi)^4} \int_{-\infty}^{\infty} \int_{-\infty}^{\infty} \frac{-1}{i\omega - Dk^2}\,e^{i(\boldsymbol{k}\cdot\boldsymbol{R} - \omega T)}\,d\boldsymbol{k}\,d\omega. \qquad (2.248)$$

In the same way as we did in one-dimensional diffusion, we first perform the integral with respect to ω:

$$I = \frac{1}{2\pi} \int_{-\infty}^{\infty} \frac{-1}{i\omega - Dk^2}\,e^{-i\omega T}\,d\omega = e^{-DTk^2},$$

for $T > 0$ and $I = 0$ for $T < 0$. We then substitute the above results for Eq. (2.248), and perform the integral with respect to $\boldsymbol{k}$ using the spherical coordinate system in that space. The method of calculation is the same as that shown in the previous subsection (§§2.4.6b), where the principal solution of the three-dimensional wave equation was sought. The result is

$$
\begin{aligned}
G &= \frac{1}{(2\pi)^3} \int_{-\infty}^{\infty} \mathrm{e}^{-DTk^2} \mathrm{e}^{\mathrm{i}\boldsymbol{k}\cdot\boldsymbol{R}} \, \mathrm{d}\boldsymbol{k} \\
&= \frac{1}{(2\pi)^3} \int_0^{2\pi} \mathrm{d}\varphi \int_0^{\pi} \sin\theta \, \mathrm{d}\theta \int_0^{\infty} k^2 \mathrm{e}^{-DTk^2 + \mathrm{i}kR\cos\theta} \, \mathrm{d}k,
\end{aligned}
$$

where we take $\boldsymbol{R}$ as a basic axis, and denote the angle between $\boldsymbol{k}$ and $\boldsymbol{R}$ by θ, and the angle of rotation around this axis by φ, so that we have $\boldsymbol{k} \cdot \boldsymbol{R} = kR\cos\theta$, $\mathrm{d}\boldsymbol{k} = k^2 \sin\theta \, \mathrm{d}k \, \mathrm{d}\theta \, \mathrm{d}\varphi$, and $R = |\boldsymbol{R}|$. The integrals with respect to φ and θ are similar to the case in which we deal with the one-dimensional diffusion, so that we obtain

$$
G = \frac{1}{2\pi^2 R} \int_0^{\infty} k \sin(kR) \mathrm{e}^{-DTk^2} \, \mathrm{d}k = \frac{1}{(4\pi DT)^{3/2}} \exp\left(-\frac{R^2}{4DT}\right),
$$
$$(2.249)$$

for $T \equiv t - \tau > 0$, and $G = 0$ for $T < 0$. ◁

We summarize the Green's functions in an infinite region in Table 2.2, where the reference equation numbers are given in the brackets as "[Eq. (⋯)]", although some of which will be derived in the next chapter.

2.5 Initial- and/or Boundary-value Problems in Rectangular Regions

To solve the problems that are given by the relevant PDEs, which satisfy the given initial and/or boundary conditions, we need to know the solutions of the homogeneous part of the respective PDEs in addition to the principal solutions mentioned above.

2.5.1 *Laplace equation*

We consider here the solution of the Laplace equation in a rectangular region ($0 \leq x \leq a$, $0 \leq y \leq b$), to which the use of the rectangular (or Cartesian) coordinate system is appropriate.

Table 2.2 Green's function in an infinite region

Here, $x = (x, y, z)$, $\boldsymbol{\xi} = (\xi, \eta, \zeta)$, $\rho = \sqrt{(x-\xi)^2 + (y-\eta)^2}$, $R = \sqrt{(x-\xi)^2 + (y-\eta)^2 + (z-\zeta)^2}$, $T = t - \tau$, $H_0^{(1)}(x)$ is the Hankel function, $K_0(x)$ is the modified Bessel function, and $1(x)$ is the unit step function. Note that the assumption of time-dependence $\exp(\pm i\omega t)$ in the wave equation (d), or the assumption of time-dependence $\exp(-\lambda t)$ in the diffusion (heat conduction) equation (e), converts the type to the Helmholtz equation (b). The change of $k \to ik$ in (b) converts to type (c). Also note the relation $H_0^{(1)}(ix) = 2/(\pi i)K_0(x)$ (see Appendix §A.2).

Type of PDE	1-dimensional	2-dimensional	3-dimensional
(a) Laplace Eq.	$(\xi - x)\,1(x-\xi)$ [Eq. (2.199)]	$-\frac{1}{2\pi}\log\rho$ [Eq. (2.206)]	$\frac{1}{4\pi R}$ [Eq. (2.213)]
(b) Helmholtz Eq.	$\frac{i}{2k}e^{ik|x-\xi|}$	$\frac{i}{4}H_0^{(1)}(k\rho)$ [Eq. (2.311)]	$\frac{1}{4\pi R}e^{ikR}$ [Eq. (2.308)]
(c) modified Helmholtz Eq.	$\frac{1}{2k}e^{-k|x-\xi|}$ [Eq. (3.70)]	$\frac{1}{2\pi}K_0(k\rho)$ [Eq. (3.193)]	$\frac{1}{4\pi R}e^{-kR}$
(d) wave Eq. [propagation to $+x$ direction]	$\frac{c}{2}\,1(cT - |x-\xi|)\,1(T)$ [Eq. (2.231)]	$\frac{c}{2\pi}\frac{1}{\sqrt{(cT)^2 - \rho^2}}\,1(cT - \rho)\,1(T)$	$\frac{c}{4\pi R}\delta(cT - R)\,1(T)$ [Eq. (2.238)]
(e) diffusion(heat conduction) Eq.	$\frac{1}{\sqrt{4\pi DT}}\exp\left(-\frac{(x-\xi)^2}{4DT}\right)1(T)$ [Eq. (2.244), (3.93)]	$\frac{1}{4\pi DT}\exp\left(-\frac{\rho^2}{4DT}\right)1(T)$	$\frac{1}{(4\pi DT)^{3/2}}\exp\left(-\frac{R^2}{4DT}\right)1(T)$ [Eq. (2.249)]

(a) $\triangle G = -\delta(\boldsymbol{x} - \boldsymbol{\xi})$,   (b) $(\triangle + k^2)\,G = -\delta(\boldsymbol{x} - \boldsymbol{\xi})$,   (c) $(\triangle - k^2)\,G = -\delta(\boldsymbol{x} - \boldsymbol{\xi})$

(d) $\left(\triangle - \frac{1}{c^2}\frac{\partial^2}{\partial t^2}\right)G = -\delta(\boldsymbol{x} - \boldsymbol{\xi})\,\delta(t - \tau)$,   (e) $\left(D\triangle - \frac{\partial}{\partial t}\right)G = -\delta(\boldsymbol{x} - \boldsymbol{\xi})\,\delta(t - \tau)$

a. Boundary-value problems in a rectangular region

In this subsection, we assume $u = 0$ on the boundary of the rectangular region $0 \leq x \leq a$, $0 \leq y \leq b$.

(i) Method of double-Fourier series expansion

Taking account of the boundary conditions, we expand the solution of $\triangle u(x, y) = 0$ in the form

$$u(x,y) = \sum_{m,n=1}^{\infty} c_{mn} \sin\left(\frac{m\pi x}{a}\right) \sin\left(\frac{n\pi y}{b}\right). \qquad (2.250)$$

We also need the Green's function $G(x, y; \xi, \eta)$, which satisfies the equation $\triangle G = -\delta(x - \xi)\,\delta(y - \eta)$. Since r.h.s. of the latter equation is expanded as

$$\delta(x - \xi) = \frac{2}{a} \sum_{m=1}^{\infty} \sin\left(\frac{m\pi x}{a}\right) \sin\left(\frac{m\pi \xi}{a}\right),$$

$$\delta(y - \eta) = \frac{2}{b} \sum_{n=1}^{\infty} \sin\left(\frac{n\pi y}{b}\right) \sin\left(\frac{n\pi \eta}{b}\right),$$

(derive them using Eqs. (2.148) and (2.151)), we may assume G of the form

$$G(x, y; \xi, \eta) = \sum_{m,n=1}^{\infty} g_{mn} \sin\left(\frac{m\pi x}{a}\right) \sin\left(\frac{n\pi y}{b}\right). \qquad (2.251)$$

After the substitution and comparison, we can determine g_{mn} so as to fulfill the given equation. By this procedure, we obtain

$$g_{mn} = \frac{4}{\pi^2 ab[(m/a)^2 + (n/b)^2]} \sin\left(\frac{m\pi \xi}{a}\right) \sin\left(\frac{n\pi \eta}{b}\right), \qquad (2.252)$$

and hence the solution

$$G(x, y; \xi, \eta) = \frac{4}{\pi^2 ab} \sum_{m,n=1}^{\infty} \frac{\sin\left(\frac{m\pi x}{a}\right) \sin\left(\frac{n\pi y}{b}\right) \sin\left(\frac{m\pi \xi}{a}\right) \sin\left(\frac{n\pi \eta}{b}\right)}{(m/a)^2 + (n/b)^2}.$$

$$(2.253)$$

(ii) Method of single-Fourier series expansion

In obtaining the Green's function, we may pay attention solely to the boundary condition at $x = 0$, a, and expand G in the form

$$G(x, y; \xi, \eta) = \frac{2}{a} \sum_{n=1}^{\infty} g_n(y, \eta) \sin\left(\frac{n\pi x}{a}\right) \sin\left(\frac{n\pi \xi}{a}\right), \qquad (2.254)$$

and substitute for the equation $\triangle G = -\delta(x - \xi)\,\delta(y - \eta)$. Then we obtain the equation for $g_n(y, \eta)$:

$$\frac{d^2 g_n}{dy^2} - \left(\frac{n\pi}{a}\right)^2 g_n = -\delta(y - \eta).$$

We also take into account that the present Green's function assumes a singularity at $y = \eta$, so that g_n has the discontinuity:

$$\left.\frac{dg_n}{dy}\right|_{\eta+0} - \left.\frac{dg_n}{dy}\right|_{\eta-0} = -1,$$

in addition to the boundary condition $g_n = 0$ at $y = 0$ and b. Under these conditions, we obtain

$$g_n = A_n \begin{cases} \sinh\left(\dfrac{n\pi y}{a}\right) \sinh\left(\dfrac{n\pi(b-\eta)}{a}\right) & (0 \le y < \eta) \\[3mm] \sinh\left(\dfrac{n\pi\eta}{a}\right) \sinh\left(\dfrac{n\pi(b-y)}{a}\right) & (\eta < y \le b) \end{cases}, \qquad (2.255)$$

where

$$A_n = \frac{a}{n\pi \sinh(n\pi b/a)}. \qquad (2.256)$$

By substitution of Eqs. (2.255) and (2.256) for Eq. (2.254) we obtain the Green's function.

The solution of the Poisson equation $\triangle u(x,y) = -f(x,y)$ in a rectangular region $(0 \le x \le a,\ 0 \le y \le b)$ with the boundary conditions $u(0,y) = u(a,y) = u(x,0) = u(x,b) = 0$ is obtained by the substitution of the Green's function (2.253) or Eqs. (2.254)–(2.256) for the following equation:

$$u(x,y) = -\int_0^a \int_0^b f(\xi,\eta) G(x,y;\xi,\eta) \, d\xi \, d\eta.$$

b. Dirichlet's problems of the Laplace equation in a rectangular region

Consider the boundary-value problem in a rectangular region D $(0 \le x \le a,\ 0 \le y \le b)$, in which $u = 0$ is given at the three sides $x = 0$, $x = a$ and $y = b$, whereas $u = g(x)$ is given at $y = 0$, i.e.,

$$u(0,y) = u(a,y) = u(x,b) = 0, \quad u(x,0) = g(x). \qquad (2.257)$$

We choose the solution of $\triangle u(x,y) = 0$ for u, and the solution of the Green's function G (Eq. (2.254) obtained in the previous subsection §§2.5.1a) for v, and apply the Green's formula (2.183). Then, the l.h.s. of the latter is

$$\iint_D [G \triangle u(x,y) - u(x,y) \triangle G] \, dx \, dy = u(\xi,\eta),$$

because $\triangle u(x,y) = 0$, and $\triangle G = -\delta(x-\xi)\,\delta(y-\eta)$ in the region D. Meanwhile, the r.h.s. of Eq. (2.183) is

$$\int_C \left(G\frac{\partial u}{\partial n} - u\frac{\partial G}{\partial n}\right) ds = \int_0^a \left(-u(x,0)\frac{\partial G}{\partial n}\right)_{y=0} dx = \int_0^a \left(g(x)\frac{\partial G}{\partial y}\right)_{y=0} dx,$$

Table 2.3 Expansion of the solution of the Laplace equation $\Delta\Phi = 0$ in cylindrical coordinate system: $\Phi(\rho, \varphi, z) = \sum_{m,k} c_{mk}\Phi_{mk}$

z dependence	Φ_{mk}
no dependence on z (2D)	$\{\rho^m, \rho^{-m}\}\{\cos m\varphi, \sin m\varphi\}$
decay as $z \to \pm\infty$	$\{J_m(k\rho), N_m(k\rho)\}\{\cos m\varphi, \sin m\varphi\}\{e^{\mp kz}\}$
wavelike as $|z| \to \infty$	$\{I_m(k\rho), K_m(k\rho)\}\{\cos m\varphi, \sin m\varphi\}\{e^{-ikz}, e^{ikz}\}$

because of the conditions $G = 0$ on the boundary C, as well as the boundary conditions (2.257) for u. (Note that the direction of n at $y = 0$ is opposite to that of dy.) Furthermore, we have

$$\left(\frac{\partial G}{\partial y}\right)_{y=0} = \frac{2}{a}\sum_{n=1}^{\infty}\sin\left(\frac{n\pi x}{a}\right)\sin\left(\frac{n\pi\xi}{a}\right)\sinh\left(\frac{n\pi(b-\eta)}{a}\right)\Big/\sinh\left(\frac{n\pi b}{a}\right).$$

Combining these results, we obtain $u(\xi, \eta)$. Thus, our final result is given by

$$u(x,y) = \frac{2}{a}\sum_{n=1}^{\infty}\left[\sin\left(\frac{n\pi x}{a}\right)\sinh\left(\frac{n\pi(b-y)}{a}\right)\Big/\sinh\left(\frac{n\pi b}{a}\right)\right]$$
$$\times\int_0^a g(\xi)\sin\left(\frac{n\pi\xi}{a}\right)\,d\xi,$$

(2.258)

where the roles of (ξ, η) and (x, y) are exchanged (see Eq. (2.223)). ◁

To obtain the solution of the Laplace equation with the boundary conditions specified on the surface of a circular cylinder, or on the surface of a sphere, it is appropriate to expand the solution in terms of the functions with cylindrical symmetry (such as the Bessel functions) or functions with spherical symmetry (such as the Legendre functions). We show in Tables 2.3 and 2.4 the basic functions necessary for these purposes. Note that the terms inside the curly brackets $\{\cdots\}$ imply to take their linear combination. The Dirichlet's problems on circular or spherical boundaries are dealt with in §§2.6.1. For further details on the Bessel functions and the Legendre functions, see e.g., [Morse and Feshbach (1953); Arfken (1985)], or Appendixes A.2 and A.3 of this book.

Table 2.4 Expansion of the solution of the Laplace equation $\triangle\Phi = 0$ in spherical coordinate system:
$$\Phi(r,\theta,\varphi) = \sum_{m,k} c_{mk} Y_{mk}$$

r dependence	Y_{mk}
decay as $r \to 0$	$\{r^n\}\{P_n^m(\cos\theta), Q_n^m(\cos\theta)\}\{\cos m\varphi, \sin m\varphi\}$
decay as $r \to \infty$	$\{r^{-(n+1)}\}\{P_n^m(\cos\theta), Q_n^m(\cos\theta)\}\{\cos m\varphi, \sin m\varphi\}$

(See also Appendix §§A.3.3 for $r = $ constant.)

2.5.2 *Wave equation*

a. Initial-value problems of one-dimensional wave equation in an infinite region

The general solution of the one-dimensional wave equation

$$\frac{\partial^2 u}{\partial x^2} = \frac{1}{c^2}\frac{\partial^2 u}{\partial t^2} \tag{2.259}$$

is given by $u(x,t) = F(x - ct)$ or $G(x + ct)$, where F and G are arbitrary functions. Here, F and G respectively describe the waves propagating in the positive and negative x directions with a speed c. If the initial conditions at $t = 0$ are given by

$$u(x,0) = f(x), \qquad \frac{\partial u(x,0)}{\partial t} = g(x), \tag{2.260}$$

the general solution is described as

$$u(x,t) = \frac{1}{2}[f(x - ct) + f(x + ct)] + \frac{1}{2c}\int_{x-ct}^{x+ct} g(x)\,\mathrm{d}x. \tag{2.261}$$

The Eq. (2.261) is called the **d'Alembert–Stokes' solution**, which may be obtained as follows:

We first assume $u(x,t) \propto \mathrm{e}^{\mathrm{i}(kx-\omega t)}$, and substitute the latter for Eq. (2.259), which yields $\omega = \pm kc$. Hence, we obtain the basic solution of $u(x,t)$ of the form $\{\mathrm{e}^{\mathrm{i}k(x+ct)}, \mathrm{e}^{\mathrm{i}k(x-ct)}\}$ or $\{\mathrm{e}^{\mathrm{i}kx}\cos(kct), \mathrm{e}^{\mathrm{i}kx}\sin(kct)\}$. Then, the general solution of u is expressed by the superposition of these solutions with arbitrary amplitudes. Taking into account that k is a continuous variable, this procedure is given by the integral of the linear combination of them with arbitrary weights (functions of k):

$$u(x,t) = \int_{-\infty}^{\infty} [A(k)\cos(kct) + B(k)\sin(kct)]\,\mathrm{e}^{\mathrm{i}kx}\,\mathrm{d}k. \tag{2.262}$$

The constants A and B are determined so as to satisfy the initial conditions:

$$u(x,0) = \int_{-\infty}^{\infty} A(k)e^{ikx}\, dk = f(x),$$

$$\frac{\partial u(x,0)}{\partial t} = \int_{-\infty}^{\infty} kcB(k)e^{ikx}\, dk = g(x),$$

from which we have[17]

$$A(k) = \frac{1}{2\pi} \int_{-\infty}^{\infty} f(x')e^{-ikx'}\, dx',$$

$$B(k) = \frac{1}{2\pi kc} \int_{-\infty}^{\infty} g(x')e^{-ikx'}\, dx'.$$

Accordingly, Eq. (2.262) becomes

$$u(x,t) = \int_{-\infty}^{\infty} \frac{1}{2\pi} \int_{-\infty}^{\infty} \left(f(x')\cos(kct) + \frac{1}{kc}g(x')\sin(kct) \right) e^{ik(x-x')}\, dx'dk$$

$$= \int_{-\infty}^{\infty} f(x') \left(\frac{1}{2\pi} \int_{-\infty}^{\infty} \cos(kct)e^{ik(x-x')}\, dk \right) dx'$$

$$+ \int_{-\infty}^{\infty} g(x') \left(\frac{1}{2\pi} \int_{-\infty}^{\infty} \frac{\sin(kct)}{kc}e^{ik(x-x')}\, dk \right) dx'.$$

Note that the integral with respect to k of the first term of r.h.s. becomes

$$\frac{1}{2\pi} \int_{-\infty}^{\infty} \frac{e^{ikct} + e^{-ikct}}{2} e^{ik(x-x')}\, dk = \frac{1}{2}[\,\delta(x - x' + ct) + \delta(x - x' - ct)\,],$$

whereas that of the second term of the r.h.s. becomes (see Eq. (2.177))

$$\frac{1}{2\pi c} \int_{-\infty}^{\infty} \frac{e^{ikct} - e^{-ikct}}{2ik} e^{ik(x-x')}\, dk = \frac{1}{2c}[\mathbf{1}(x - x' + ct) - \mathbf{1}(x - x' - ct)],$$

so that we can derive the Eq. (2.261):

$$u(x,t) = \frac{1}{2}[f(x - ct) + f(x + ct)] + \frac{1}{2c} \int_{x-ct}^{x+ct} g(x)\, dx. \qquad (2.261)$$

---

[17]Based on the knowledge thus far stated, $A(k)$, for instance, is determined as follows. We integrate both sides of the first condition:

$$\frac{1}{2\pi} \int_{-\infty}^{\infty} [\cdots]e^{-ikx'} dx' \quad \leftarrow \quad \left[\int_{-\infty}^{\infty} A(k')e^{ik'x'}\, dk' = f(x') \right].$$

On the l.h.s., we exchange the order of integration with respect to x' and k', and use the relation of the delta function Eq. (2.176). Similarly, $B(k)$ is calculated. The present results are a direct consequence of the Fourier integral formula (3.39) shown later.

For the interpretation of the r.h.s. of the above equation, readers may refer to the characteristic curves mentioned in §§2.2.2. In particular, the second term reflects the "region of influence". Note that the method of solution is also dealt with later in the example of the Fourier transforms (§§3.3.2).[18]

b. Initial-value problems of three-dimensional wave equation in an infinite region

We consider here the three-dimensional wave equation

$$\Delta u = \frac{1}{c^2} \frac{\partial^2 u}{\partial t^2}, \tag{2.263}$$

under the initial conditions

$$u(\boldsymbol{x}, 0) = f(\boldsymbol{x}), \qquad \frac{\partial u(\boldsymbol{x}, 0)}{\partial t} = g(\boldsymbol{x}). \tag{2.264}$$

In the same manner as is made in a one-dimensional case, we assume the basic solution of the form $u(\boldsymbol{x}, t) \propto e^{i(\boldsymbol{k} \cdot \boldsymbol{x} - \omega t)}$, and substitute for

---

[18]The solution may also be obtained as follows: By the transformation of variables from (x, y) to (ξ, η), such that $\xi = x - ct$, $\eta = x + ct$, Eq. (2.259) becomes

$$\frac{\partial^2 u}{\partial \xi \partial \eta} = 0,$$

which is satisfied by $u = F(\xi)$ or $u = G(\eta)$ (F, G being arbitrary functions). Then, the general solution of Eq. (2.259) is given by $u = F(x - ct) + G(x + ct)$. Imposing the initial conditions, we have

$$u(x, 0) = F(x) + G(x) = f(x), \tag{*}$$

and

$$\frac{\partial u(x, 0)}{\partial t} = -cF'(x) + cG'(x) = g(x).$$

If we integrate the latter with respect to x, we have

$$F(x) - G(x) = -\frac{1}{c} \int_{-\infty}^{x} g(x') \, dx', \tag{**}$$

By the addition and/or subtraction of Eq. (*) and Eq. (**), we have

$$F(x) = \frac{1}{2} f(x) - \frac{1}{2c} \int_{-\infty}^{x} g(x') \, dx', \qquad G(x) = \frac{1}{2} f(x) + \frac{1}{2c} \int_{-\infty}^{x} g(x') \, dx'.$$

Consequently, the general solution $u = F(x - ct) + G(x + ct)$ becomes

$$u(x, t) = \frac{1}{2} f(x - ct) - \frac{1}{2c} \int_{-\infty}^{x-ct} g(x') \, dx' + \frac{1}{2} f(x + ct) + \frac{1}{2c} \int_{-\infty}^{x+ct} g(x') \, dx'$$

$$= \frac{1}{2} [f(x - ct) + f(x + ct)] + \frac{1}{2c} \int_{x-ct}^{x+ct} g(x') \, dx'.$$

Eq. (2.263), which yields $\omega = \pm kc$. Hence, we have the basic solution $u(\boldsymbol{x}, t) = \{e^{i\boldsymbol{k}\cdot\boldsymbol{x}}\cos(kct), e^{i\boldsymbol{k}\cdot\boldsymbol{x}}\sin(kct)\}$, where $\boldsymbol{x} = (x, y, z)$, $\boldsymbol{k} = (k_x, k_y, k_z)$ and $k = |\boldsymbol{k}| = \sqrt{k_x^2 + k_y^2 + k_z^2}$.

The general solution of u is obtained by the integral of the above-mentioned basic solutions with the magnitudes depending on $\boldsymbol{k}$, so that we have

$$u(\boldsymbol{x}, t) = \int_{-\infty}^{\infty} [A(\boldsymbol{k})\cos(kct) + B(\boldsymbol{k})\sin(kct)]e^{i\boldsymbol{k}\cdot\boldsymbol{x}} \, d\boldsymbol{k}. \qquad (2.265)$$

The constants A and B are determined so as to satisfy the initial conditions:

$$u(\boldsymbol{x}, 0) = \int_{-\infty}^{\infty} A(\boldsymbol{k})e^{i\boldsymbol{k}\cdot\boldsymbol{x}} \, d\boldsymbol{k} = f(\boldsymbol{x}),$$

$$\frac{\partial u(\boldsymbol{x}, 0)}{\partial t} = \int_{-\infty}^{\infty} kcB(\boldsymbol{k})e^{i\boldsymbol{k}\cdot\boldsymbol{x}} \, d\boldsymbol{k} = g(\boldsymbol{x}),$$

from which we have

$$A(\boldsymbol{k}) = \frac{1}{(2\pi)^3} \int_{-\infty}^{\infty} f(\boldsymbol{x}')e^{-i\boldsymbol{k}\cdot\boldsymbol{x}'} \, d\boldsymbol{x}',$$

$$B(\boldsymbol{k}) = \frac{1}{(2\pi)^3 kc} \int_{-\infty}^{\infty} g(\boldsymbol{x}')e^{-i\boldsymbol{k}\cdot\boldsymbol{x}'} \, d\boldsymbol{x}'.$$

Accordingly, we have

$$u(\boldsymbol{x}, t) = \int_{-\infty}^{\infty} f(\boldsymbol{x}') \left(\frac{1}{(2\pi)^3} \int_{-\infty}^{\infty} \cos(kct)e^{i\boldsymbol{k}\cdot(\boldsymbol{x}-\boldsymbol{x}')} \, d\boldsymbol{k} \right) d\boldsymbol{x}'$$

$$+ \int_{-\infty}^{\infty} g(\boldsymbol{x}') \left(\frac{1}{(2\pi)^3} \int_{-\infty}^{\infty} \frac{\sin(kct)}{kc}e^{i\boldsymbol{k}\cdot(\boldsymbol{x}-\boldsymbol{x}')} \, d\boldsymbol{k} \right) d\boldsymbol{x}'.$$

The integral with respect to $\boldsymbol{k}$ is performed over the entire $\boldsymbol{k}$ space, which is calculated in the spherical coordinate system in that space. The calculation is basically the same as the one made at §§2.4.6b. Namely, we choose $\boldsymbol{R} = \boldsymbol{x} - \boldsymbol{x}'$ as the basic axis, and denote the angle between $\boldsymbol{k}$ and $\boldsymbol{R}$ by θ, the angle around this axis by φ, and $R = |\boldsymbol{R}|$. Then, the integral of $\boldsymbol{k}$ in the second term of the r.h.s. becomes

$$I = \frac{1}{(2\pi)^3} \int_{-\infty}^{\infty} \frac{\sin(kct)}{kc} e^{i\boldsymbol{k}\cdot\boldsymbol{R}} \, d\boldsymbol{k} = \ldots = \frac{1}{4\pi cR}\delta(R - ct).$$

The derivation of the last expression is the same as that Eqs. (2.236)–(2.238) except for the proportional constant c^2. Similarly, the integral of $\boldsymbol{k}$ in the first term of the r.h.s. becomes

$$J = \frac{1}{(2\pi)^3} \int_{-\infty}^{\infty} \cos(kct) \, e^{i\boldsymbol{k}\cdot\boldsymbol{R}} \, d\boldsymbol{k}$$

$$= \frac{1}{(2\pi)^3} \int_0^{\infty} dk \, k^2 \int_0^{\pi} d\theta \, \sin\theta \int_0^{2\pi} d\varphi \, \cos(kct) \, e^{ikR\cos\theta}.$$

The integral with respect to φ results in a multiplication by 2π. As for the integral with respect to θ, we put $\cos\theta = x$ and obtain

$$\int_0^\pi d\theta\, \sin\theta\, e^{ikR\cos\theta} = \int_{-1}^1 e^{ikRx}\, dx = \frac{e^{ikR} - e^{-ikR}}{ikR},$$

which is the same as before. Therefore, for $t > 0$, we have

$$J = \frac{1}{(2\pi)^2}\int_0^\infty k\cos(kct)\frac{e^{ikR} - e^{-ikR}}{iR}\, dk = \ldots^{19} = \frac{1}{4\pi cR}\frac{\partial}{\partial t}\delta(R - ct).$$

Consequently, we have

$$u(\boldsymbol{x}, t) = \frac{1}{4\pi c}\int_{-\infty}^\infty \frac{1}{R}\left(f(\boldsymbol{x}')\frac{\partial}{\partial t}\delta(R - ct) + g(\boldsymbol{x}')\delta(R - ct)\right)\, d\boldsymbol{x}'. \quad (2.266)$$

From a view point of physics, the wave observed at a point $\boldsymbol{x}$ at time t is a superposition of the wave source elements at a distance ct from this point, which is called the **Huygens' principle**.[20]

2.5.3 *Diffusion equation*

We learned in §§2.4.7a that the principal solution of the one-dimensional diffusion equation or heat conduction equation Eq. (2.239):

$$L[G] \equiv D\frac{\partial^2 G}{\partial x^2} - \frac{\partial G}{\partial t} = -\delta(x - \xi)\,\delta(t - \tau) \quad (2.267)$$

---

[19]The direct calculation is as follows:

$$J = \ldots = \frac{1}{(2\pi)^2 icR}\frac{\partial}{\partial t}\int_0^\infty \sin(kct)\,(e^{ikR} - e^{-ikR})\, dk$$

$$= -\frac{1}{8\pi cR}\frac{\partial}{\partial t}\left(\frac{1}{2\pi}\int_{-\infty}^\infty (e^{ikct} - e^{-ikct})\,(e^{ikR} - e^{-ikR})\, dk\right)$$

$$= -\frac{1}{4\pi cR}\frac{\partial}{\partial t}[\,\delta(R + ct) - \delta(R - ct)\,] = \frac{1}{4\pi cR}\frac{\partial}{\partial t}\delta(R - ct),$$

where the parity of the delta function $\delta(-x) = \delta(x)$ and the property $\delta(x) = 0$ for $x > 0$ are taken into account.

[20]For further calculation, the following property of $\delta(x)$ may be useful:

$$\int_{-\infty}^\infty f(\xi)\,\delta'(\xi - x)\, d\xi = -f'(x).$$

In terms of a local spherical coordinates (R, θ', φ') with the origin at $\boldsymbol{x}$, and expressing $f(\boldsymbol{x}')$ and $g(\boldsymbol{x}')$ such as $f(R, \theta', \varphi')$ and $g(R, \theta', \varphi')$, $d\boldsymbol{x}' = R^2 dR\, d\Omega$, where $d\Omega = \sin\theta'\, d\theta'\, d\varphi'$ is a spherical surface element, we can rewrite Eq. (2.266) as

$$u(\boldsymbol{x}, t) = \frac{1}{4\pi}\int\left[\frac{\partial}{\partial t}(t\,f(ct, \theta', \varphi')) + t\,g(ct, \theta', \varphi')\right]\, d\Omega(\theta', \varphi').$$

is given by Eq. (2.244):

$$G(x,t;\xi,\tau) = \frac{1}{\sqrt{4\pi D(t-\tau)}}\exp\left(-\frac{(x-\xi)^2}{4D(t-\tau)}\right), \quad (t > \tau), \quad (2.268)$$

and $G = 0$ for $t < \tau$, where D is a positive constant called the diffusion constant, or thermal diffusivity. On the basis of these results, we consider the initial- and/or boundary-value problems.

a. Initial-value problems in an infinite region

We shall obtain the initial-value problem of the one-dimensional diffusion equation, in which the following condition

$$u(x,0) = f(x) \tag{2.269}$$

is given in the range $-\infty < x < \infty$ at $t = 0$, by the use of the generalized Green's formula (2.182).

With reference to Eq. (2.179), the adjoint differential operator M, together with P and Q, corresponding to the differential operator L, is given by putting $y = t$, $A = D$ (here D is the diffusion constant) and $E = -1$ (where other coefficients are zero):

$$M[v] = D\frac{\partial^2 v}{\partial x^2} + \frac{\partial v}{\partial t}, \qquad P = D\left(v\frac{\partial u}{\partial x} - u\frac{\partial v}{\partial x}\right), \qquad Q = -uv,$$

so that Eq. (2.182) becomes

$$\int_S (vL[u] - uM[v])\ \mathrm{d}x\,\mathrm{d}t = \int_C \left[D\left(v\frac{\partial u}{\partial x} - u\frac{\partial v}{\partial x}\right)\,\mathrm{d}t + uv\,\mathrm{d}x\right]. \tag{2.270}$$

If we choose H for v, where H belongs to M, such that

$$M[H] \equiv D\frac{\partial^2 H}{\partial x^2} + \frac{\partial H}{\partial t} = -\delta(x-\xi)\,\delta(t-\tau), \tag{2.271}$$

whereas $L[u] = 0$, then Eq. (2.270) becomes

$$u(\xi,\tau) = \int_C \left[D\left(H(x,t;\xi,\tau)\frac{\partial u(x,t)}{\partial x} - u(x,t)\frac{\partial H(x,t;\xi,\tau)}{\partial x}\right)\,\mathrm{d}t\right.$$

$$\left. + u(x,t)H(x,t;\xi,\tau)\,\mathrm{d}x\right], \tag{2.272}$$

where C is the path surrounding the rectangular region S, *i.e.*, $P_1P_2P_3P_4P_1$ as shown in Fig. 2.15(a), and (ξ, τ) is an inner point of region S. Taking account of the reciprocity principle (2.222):

$$H(x,t;\xi,\tau) = G(\xi,\tau;x,t)$$

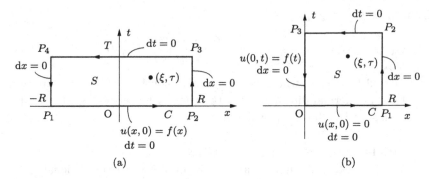

Fig. 2.15 Application of the Green's function: (a) initial-value problem, and (b) initial-
and boundary-value problem.

and Eq. (2.244) [reproduced in Eq. (2.268)], the solution of Eq. (2.271) is
given by

$$H(x,t;\xi,\tau) = \frac{1}{\sqrt{4\pi D(\tau - t)}} \exp\left(-\frac{(\xi - x)^2}{4D(\tau - t)}\right), \quad (\tau > t),$$

$$H(x,t;\xi,\tau) = 0, \quad (\tau < t).$$

(2.273)

For further calculation of Eq. (2.272), we note that the integrand is

$P_1P_2 \ (-R \le x \le R):$ $dt = 0,$ and $u = u(x,0) = f(x),$

$P_2P_3:$ $dx = 0,$ and $H, \partial H/\partial x \to 0$ as $R \to \infty,$

$P_3P_4:$ $dt = 0,$ and $H = 0$ because $t > \tau,$

$P_4P_1:$ $dx = 0,$ and $H, \partial H/\partial x \to 0$ as $R \to \infty,$

from which we finally have

$$u(\xi,\tau) = \int_{P_1}^{P_2} u(x,0)H(x,0;\xi,\tau)\,dx$$

$$\to \int_{-\infty}^{\infty} \frac{f(x)}{\sqrt{4\pi D\tau}} \exp\left(-\frac{(\xi - x)^2}{4D\tau}\right)\,dx \quad (R \to \infty).$$

Exchanging the roles of (x,t) and (ξ,η), we have

$$u(x,t) = \frac{1}{\sqrt{4\pi Dt}} \int_{-\infty}^{\infty} f(\xi)\,\exp\left(-\frac{(x - \xi)^2}{4Dt}\right)\,d\xi.$$

(2.274)

A further transformation of the variable from ξ to η in Eq. (2.274), such
that $\xi = x + 2\sqrt{Dt}\,\eta$, yields

$$u(x,t) = \frac{1}{\sqrt{\pi}} \int_{-\infty}^{\infty} f(x + 2\sqrt{Dt}\,\eta)\,\exp(-\eta^2)\,d\eta.$$

(2.275)

b. Initial- and boundary-value problems in a semi-infinite region

We now consider a one-dimensional heat conduction (or diffusion) in a semi-infinite space $0 \leq x < \infty$. We assume that $u = 0$ at all locations at an initial instant $(t = 0)$, and that $u = f(t)$ is applied at the left end of the region $x = 0$:

$$u(x, 0) = 0, \qquad u(0, t) = f(t); \qquad u(x, t) \to 0 \quad \text{as} \quad x \to \infty.$$

We have previously obtained the principal solution of the one-dimensional diffusion equation (2.244):

$$G(x, t; \xi, \tau) = \frac{1}{\sqrt{4\pi D(t - \tau)}} \exp\left(-\frac{(x - \xi)^2}{4D(t - \tau)}\right) \quad \text{for } t > \tau,$$

and $G = 0$ for $t < \tau$, which shows the development of the density distribution of a material (or heat) that is concentrated at $x = \xi$ at time $t = \tau$ with its total amount of unity.

Making use of the above expression, we assume the Green's function valid for $t > \tau$:

$$G_1(x, t; \xi, \tau) = \frac{1}{\sqrt{4\pi D(t - \tau)}} \left[\exp\left(-\frac{(x - \xi)^2}{4D(t - \tau)}\right) - \exp\left(-\frac{(x + \xi)^2}{4D(t - \tau)}\right)\right],$$
(2.276)

and $G_1 = 0$ for $t < \tau$. As is easily observed from the symmetry, $G_1 = 0$ always holds at $x = 0$. In a similar manner as before (§§2.5.3a), we define the adjoint Green's function H_1 that corresponds to Eq. (2.276):

$$H_1(x, t; \xi, \tau) = \frac{1}{\sqrt{4\pi D(\tau - t)}} \left[\exp\left(-\frac{(\xi - x)^2}{4D(\tau - t)}\right) - \exp\left(-\frac{(\xi + x)^2}{4D(\tau - t)}\right)\right],$$
(2.277)

which is valid for $\tau > t$, and $H_1 = 0$ for $\tau < t$. If we choose H_1 for v of Eq. (2.270), whereas u is a solution of $L[u] = 0$, we have

$$u(\xi, \tau) = \int_C D\left(H_1(x, t; \xi, \tau)\frac{\partial u(x, t)}{\partial x} - u(x, t)\frac{\partial H_1(x, t; \xi, \tau)}{\partial x}\right) dt$$

$$+ \int_C u(x, t)H_1(x, t; \xi, \tau)\, dx.$$

Here, C is the path surrounding the rectangular region S, *i.e.*, $\mathrm{OP_1P_2P_3O}$, as shown in Fig. 2.15(b), and (ξ, τ) is an inner point of that region. For the calculation of the above integral, we note that the integrand is

$$\mathrm{OP_1} \ (0 \leq x \leq R) : \mathrm{d}t = 0, \quad \text{and} \quad u = u(x, 0) = 0,$$
$$\mathrm{P_1P_2} : \quad \mathrm{d}x = 0, \quad \text{and} \quad H_1, \ \partial H_1/\partial x \to 0 \quad \text{as} \quad R \to \infty,$$
$$\mathrm{P_2P_3} : \quad \mathrm{d}t = 0, \quad \text{and} \quad H_1 = 0 \quad \text{because} \quad t > \tau,$$

$$\mathrm{P_3O}: \quad \mathrm{d}x = 0, \quad H_1 = 0 \quad \text{and} \quad u(0,t) = f(t),$$

from which we have

$$u(\xi,\tau) = -\int_{P_3}^{O} D\, u(0,t)\frac{\partial}{\partial x}H_1(0,t;\xi,\tau)\,\mathrm{d}t$$

$$= \frac{D\xi}{2\sqrt{\pi}}\int_0^\tau \frac{f(t)}{[D(\tau-t)]^{3/2}}\exp\left(-\frac{\xi^2}{4D(\tau-t)}\right)\mathrm{d}t.$$

Exchanging the roles of (x,t) and (ξ,η), we have

$$u(x,t) = \frac{Dx}{2\sqrt{\pi}}\int_0^t \frac{f(\tau)}{[D(t-\tau)]^{3/2}}\exp\left(-\frac{x^2}{4D(t-\tau)}\right)\mathrm{d}\tau. \tag{2.278}$$

In terms of the variable $x^2/[4D(t-\tau)] = \eta^2$, Eq. (2.278) becomes

$$u(x,t) = \frac{2}{\sqrt{\pi}}\int_{x/\sqrt{4Dt}}^\infty f\left(t - \frac{x^2}{4D\eta^2}\right)\exp(-\eta^2)\,\mathrm{d}\eta. \tag{2.279}$$

◁

In particular, if $f = T_0$ (constant), Eq. (2.279) becomes

$$u(x,t) = T_0\frac{2}{\sqrt{\pi}}\int_{x/\sqrt{4Dt}}^\infty \exp(-\eta^2)\,\mathrm{d}\eta = T_0\,\mathrm{erfc}\left(\frac{x}{\sqrt{4Dt}}\right), \tag{2.280}$$

where $\mathrm{erfc}(x)$ is called the **complementary error function**, which, together with the **error function** $\mathrm{erf}(x)$, is defined as

$$\mathrm{erf}(x) = \frac{2}{\sqrt{\pi}}\int_0^x e^{-\eta^2}\,\mathrm{d}\eta, \qquad \mathrm{erfc}(x) = 1 - \mathrm{erf}(x). \tag{2.281}$$

The error function describes the integral of the normalized Gaussian distribution in the interval $[0,x]$, which is depicted in Fig. 2.16.

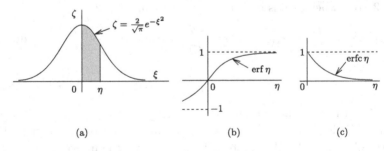

(a)   (b)   (c)

Fig. 2.16 (a) Gaussian distribution, (b) error function, and (c) complementary error function.

2.6 Initial- and/or Boundary-value Problems in Circular or Spherical Regions

2.6.1 *Dirichlet's problems for circular or spherical regions*

In this subsection, we consider the Dirichlet's problems for circular or spherical regions.

a. Dirichlet's problem on the potential in a circular region

We use here both r and ρ. To avoid confusion, we employ r as one of the polar coordinates (r, θ). Our purpose is to obtain a solution of the potential problem (*i.e.*, the solution of the Laplace equation), in which the boundary value is given on a circle of radius a, *i.e.*,

$$\triangle u(r, \theta) = 0 \quad \text{in the circular region } D \ (r \leq a),$$
$$u(a, \theta) = f(\theta) \quad \text{on the circular boundary } C.$$

To do so, we make use of the Green's formula (2.183):

$$\int\int_D (v \triangle u - u \triangle v)\, dx\, dy = \int_C \left(v \frac{\partial u}{\partial n} - u \frac{\partial v}{\partial n} \right) ds.$$

If we choose G for v, such that $\triangle G(\boldsymbol{x}, \boldsymbol{x}') = -\delta(\boldsymbol{x} - \boldsymbol{x}')$ and $G = 0$ on the boundary C ($r = a$), as well as u such that $\triangle u = 0$, then we will be able to obtain the solution

$$u(\boldsymbol{x}') = \int_C \left(-u(\boldsymbol{x}) \frac{\partial G(\boldsymbol{x}, \boldsymbol{x}')}{\partial n} \right) ds(\boldsymbol{x}).$$

Our first task is to find G that satisfies the above-mentioned requirements. We know that the regular solution of

$$\triangle u(\boldsymbol{x}) \equiv \frac{\partial^2 u}{\partial r^2} + \frac{1}{r} \frac{\partial u}{\partial r} + \frac{1}{r^2} \frac{\partial^2 u}{\partial \theta^2} = 0$$

is given in the form $u(r, \theta) = R(r)\Theta(\theta)$. From the physical requirement that $u(r, \theta)$ must be a smooth and one-valued periodic function of θ with period 2π, we have $\Theta(\theta) = \{\cos(n\theta), \sin(n\theta)\}$ (n : integer), and also taking account of the regularity at $r = 0$, we have

$$u(r, \theta) = \frac{\tilde{a}_0}{2} + \sum_{n=1}^{\infty} r^n [\tilde{a}_n \cos(n\theta) + \tilde{b}_n \sin(n\theta)],$$

where $\tilde{a}_n$, $\tilde{b}_n$ are arbitrary constants. Meanwhile, the principal solution of the two-dimensional Laplace equation is given by (2.206) [or (2.216)]:

$$G_0(\boldsymbol{x}, \boldsymbol{x}') = -\frac{1}{2\pi} \log \rho = \frac{1}{2\pi} \log \frac{1}{|\boldsymbol{x} - \boldsymbol{x}'|}, \qquad \rho = |\boldsymbol{x} - \boldsymbol{x}'|,$$

so that $G = G_0 +$ (regular solution of the Laplace equation). To impose the boundary condition at $r = a$, we express the position $\boldsymbol{x}$ (point P) and $\boldsymbol{x}'$ (point Q) in a polar coordinate system, *i.e.*, $\boldsymbol{x} = (r, \theta)$ and $\boldsymbol{x}' = (r', \theta')$ (see Fig. 2.17). For brevity, we denote the relative angle by $\varphi = \theta - \theta'$, and the distance between P and Q by $\rho = \sqrt{r^2 - 2rr'\cos\varphi + r'^2}$. The latter is rewritten for $r' < r$ as[21]

$$\log \rho = \log \sqrt{r^2 - 2rr'\cos\varphi + r'^2} = \log r - \sum_{n=1}^{\infty} \frac{1}{n}\left(\frac{r'}{r}\right)^n \cos(n\varphi),$$

so that we have, after considering the redundancy of the regular solution,

$$G(\boldsymbol{x}, \boldsymbol{x}') = -\frac{1}{2\pi}\left[\log r - \sum_{n=1}^{\infty} \frac{1}{n}\left(\frac{r'}{r}\right)^n \cos(n\varphi)\right]$$
$$+ \frac{a_0}{2} + \sum_{n=1}^{\infty} r^n [a_n \cos(n\varphi) + b_n \sin(n\varphi)].$$

The boundary condition $G = 0$ at $r = a$ requires

$$a_0 = \frac{1}{\pi}\log a, \qquad a_n = -\frac{1}{2\pi n}\left(\frac{r'}{a^2}\right)^n, \qquad b_n = 0,$$

so that we obtain

$$G(\boldsymbol{x}, \boldsymbol{x}') = \frac{1}{2\pi}\left\{\log\left(\frac{a}{r}\right) - \sum_{n=1}^{\infty} \frac{1}{n}\left[\left(\frac{r'}{a^2}\right)^n r^n - \left(\frac{r'}{r}\right)^n\right]\cos(n\varphi)\right\}.$$
$$(2.282)$$

Before proceeding further, consider the point Q^*, such that it satisfies

$$r'/a^2 = 1/r'', \qquad i.e., \qquad r'r'' = a^2. \qquad (2.283)$$

---

[21] Note that

$$\log \sqrt{r^2 - 2rr'\cos\varphi + r'^2} = \log r + \frac{1}{2}\log(1 - 2\zeta\cos\varphi + \zeta^2)$$
$$= \log r + \frac{1}{2}\log(1 - \zeta e^{i\varphi})(1 - \zeta e^{-i\varphi}), \quad (\zeta \equiv r'/r).$$

By using an expansion valid for $|\epsilon| < 1$: $\log(1 - \epsilon) = -\sum_{n=1}^{\infty} \epsilon^n/n$, we have

$$\log \sqrt{r^2 - 2rr'\cos\varphi + r'^2} = \log r - \frac{1}{2}\sum_{n=1}^{\infty} \frac{1}{n}\left[(\zeta e^{i\varphi})^n + (\zeta e^{-i\varphi})^n\right]$$

$$= \log r - \sum_{n=1}^{\infty} \frac{1}{n}\zeta^n \cos(n\varphi) = \log r - \sum_{n=1}^{\infty} \frac{1}{n}\left(\frac{r'}{r}\right)^n \cos(n\varphi).$$

Then, the expression of the first term in $[\cdots]$ of Eq. (2.282) is rewritten as

$$-\sum_{n=1}^{\infty} \frac{1}{n} \left(\frac{r'}{a^2}\right)^n r^n \cos(n\varphi) = -\sum_{n=1}^{\infty} \frac{1}{n} \left(\frac{r}{r''}\right)^n \cos(n\varphi)$$

$$= \log \sqrt{r''^2 - 2r''r\cos\varphi + r^2} - \log r''$$

$$= \log \rho^* - \log r'',$$

where $\rho^* = \sqrt{r''^2 - 2r''r\cos\varphi + r^2}$. Accordingly, Eq. (2.282) becomes

$$G = \frac{1}{2\pi} \log\left(\frac{a}{r}\frac{\rho^*}{r''}\frac{r}{\rho}\right) = \frac{1}{2\pi} \log\left(\frac{a\,\rho^*}{r''\rho}\right) = \frac{1}{2\pi} \log\left(\frac{r'\rho^*}{a\,\rho}\right). \qquad (2.284)$$

From a geometrical perspective, point Q^* is the **mirror image** of point Q with respect to a circle of radius a. Namely, as shown in Fig. 2.17, point Q^* is situated on the line passing through points O and Q with distance $r''\ (= a^2/r')$ from point O, which gives the relation $a/r' = r''/a$. If point P is on the circle ($r = a$), then the triangle OPQ and the triangle OQ^*P are similar in shape (the angle φ between the two sides that meet at point O is common to both). It then follows that $a/r' = \rho^*/\rho$, and hence $G = 0$ as is understood from Eq. (2.284), *i.e.*, which shows that G satisfies the boundary condition on the circle. ◁

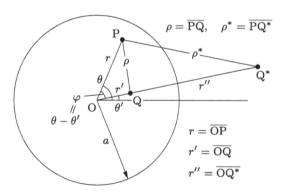

Fig. 2.17   Mirror image for the circle.

Using the Green's function G defined by Eq. (2.284), we can obtain the solution of the potential problem. Namely, we substitute $G = 1/(2\pi) \log(r'\rho^*/a\rho)$, with

$$\rho = \sqrt{r^2 - 2rr'\cos(\theta - \theta') + r'^2}, \quad \rho^* = \sqrt{r^2 - 2rr''\cos(\theta - \theta') + r''^2},$$

as well as the boundary value $u(a, \theta) = f(\theta)$ for the Green's formula mentioned in the present subsection:

$$u(r', \theta') = \int_C \left(-u \frac{\partial G}{\partial n} \right) ds = - \int_0^{2\pi} f(\theta) \left(\frac{\partial G}{\partial r} \right)_{r=a} a \, d\theta.$$

After some calculations,[22] we have

$$\left(\frac{\partial G}{\partial r} \right)_{r=a} = -\frac{1}{2\pi a} \cdot \frac{a^2 - r'^2}{a^2 - 2ar' \cos(\theta - \theta') + r'^2},$$

so that our final solution is

$$u(r, \theta) = \frac{1}{2\pi} \int_0^{2\pi} \frac{(a^2 - r^2) f(\theta')}{a^2 - 2ar \cos(\theta' - \theta) + r^2} \, d\theta', \qquad (2.285)$$

where we have exchanged the roles of (r', θ') and (r, θ). The equation (2.285) is called the **Poisson integral formula**. If Eq. (2.285) is rewritten in the form

$$u(r, \theta) = \int_0^{2\pi} f(\theta') \, G(r, \theta; a, \theta') \, d\theta', \quad G = \frac{1}{2\pi} \frac{a^2 - r^2}{a^2 - 2ar \cos(\theta' - \theta) + r^2},$$

then G is regarded as the Green's function of the present boundary-value problem. The derivation of the Poisson's formula is also given in a complex function theory (see, *e.g.*, [Morse and Feshbach (1953); Carrier, Krook and Pearson (1966); Brown and Churchill (2009); Fujiwara (2013)], *etc.*).

b. Dirichlet's problem on the potential in a spherical region

Similar to the previous subsection, we construct the solution G that fulfills the condition $G = 0$ on a spherical surface of radius $r = a$.

22

$$2\pi \left(\frac{\partial G}{\partial r} \right)_{r=a} = \frac{1}{2} \left(\frac{\partial}{\partial r} \log \left(\frac{r' \rho *}{a \rho} \right)^2 \right)_{r=a}$$

$$= \left(\frac{r - r'' \cos(\theta - \theta')}{r^2 - 2rr'' \cos(\theta - \theta') + r''^2} - \frac{r - r' \cos(\theta - \theta')}{r^2 - 2rr' \cos(\theta - \theta') + r'^2} \right)_{r=a}$$

$$= \frac{a - r'' \cos(\theta - \theta')}{a^2 - 2ar'' \cos(\theta - \theta') + r''^2} - \frac{a - r' \cos(\theta - \theta')}{a^2 - 2ar' \cos(\theta - \theta') + r'^2}. \qquad (*)$$

By the substitution of $r'' = a^2/r'$, the first term of the r.h.s. becomes

$$\frac{a - r'' \cos(\theta - \theta')}{a^2 - 2ar'' \cos(\theta - \theta') + r''^2} = \frac{r'}{a} \frac{r' - a \cos(\theta - \theta')}{a^2 - 2ar' \cos(\theta - \theta') + r'^2},$$

which is put together with the second term of the r.h.s. of Eq. (*).

Using the principal solution of the three-dimensional Laplace equation (2.213) [or (2.218)]:

$$G_0(\boldsymbol{x}, \boldsymbol{x}') = \frac{1}{4\pi} \frac{1}{|\boldsymbol{x} - \boldsymbol{x}'|},$$

and the regular solution u of the Laplace equation $\triangle u(\boldsymbol{x}) = 0$, we shall find $G \equiv G_0 + u$, such that $G = 0$ is satisfied at $r = a$. Then, we apply the Green's formula to solve the boundary-value problems that satisfy the condition specified on the spherical boundary.

Considering the present geometry of the boundary, it is appropriate to seek the solution in the spherical coordinate system. We first employ the solution valid in the region $r < a$:

$$u(r, \theta, \varphi) = \sum_{n=0}^{\infty} a_n r^n P_n(\cos \omega),$$

where P_n is the Legendre function (see Eq. (A.49) in §§A.3.1). We denote the position vector of point P by $\boldsymbol{x}$ and point Q by $\boldsymbol{x}'$, which are given by (r, θ, φ) and (r', θ', φ'), respectively, in the spherical coordinate system. Thus, the distance between these two points is $R = |\boldsymbol{x} - \boldsymbol{x}'| = \sqrt{r^2 - 2rr' \cos \omega + r'^2}$, where ω is the angle between the two vectors that satisfies $\cos \omega = \cos \theta \cos \theta' + \sin \theta \sin \theta' \cos(\varphi - \varphi')$. By combining these solutions, we can construct the Green's function

$$G = \frac{1}{4\pi \sqrt{r^2 - 2rr' \cos \omega + r'^2}} + \sum_{n=0}^{\infty} a_n r^n P_n(\cos \omega),$$

where the constants a_n are determined so as to fulfill the condition $G = 0$ at $r = a$. To do this, we make use of the generating function of the Legendre function (A.59):

$$\frac{1}{R} = \frac{1}{\sqrt{r^2 - 2rr' \cos \omega + r'^2}} = \sum_{n=0}^{\infty} \frac{r'^n}{r^{n+1}} P_n(\cos \omega), \qquad (r' < r),$$

which is substituted for the above equation to determine a_n,[23] so that we obtain

$$G(r, \theta, \varphi; r', \theta', \varphi') = \frac{1}{4\pi} \sum_{n=0}^{\infty} \left(\frac{r'^n}{r^{n+1}} - \frac{r'^n r^n}{a^{2n+1}} \right) P_n(\cos \omega). \qquad (2.286)$$

---

[23]The equation for G is rewritten as

$$G = \sum_{n=0}^{\infty} \left(\frac{1}{4\pi} \frac{r'^n}{r^{n+1}} + a_n r^n \right) P_n(\cos \omega).$$

By imposing the condition $G = 0$ at $r = a$, we have $a_n = -r'^n/(4\pi a^{2n+1})$.

In the same manner as we dealt with previously (§§2.6.1a), we introduce a **mirror image** point Q^* (position vector $\boldsymbol{x}''$), which is on the line passing through two points O and Q with a distance $r''(= a^2/r')$ between O and Q^*. Using this point, Eq. (2.286) is rewritten as

$$
\begin{aligned}
G(r,\theta,\varphi;\ r',\theta',\varphi') &= \frac{1}{4\pi}\sum_{n=0}^{\infty}\left(\frac{r'^{\,n}}{r^{n+1}} - \frac{r''}{a}\frac{r^n}{r''^{\,n+1}}\right)P_n(\cos\omega) \\
&= \frac{1}{4\pi}\left(\frac{1}{R} - \frac{r''}{a}\frac{1}{R^*}\right) = \frac{1}{4\pi}\left(\frac{1}{R} - \frac{a}{r'}\frac{1}{R^*}\right),
\end{aligned}
\tag{2.287}
$$

where $R^* = |\boldsymbol{x} - \boldsymbol{x}''| = \sqrt{r^2 - 2rr''\cos\omega + r''^{\,2}}$.   ◁

We now obtain the solution of the potential problem, in which the boundary value is specified on the sphere of radius a, *i.e.*,

$$\triangle u(r,\theta,\varphi) = 0 \quad \text{in the spherical region}\quad r \le a,$$
$$u(a,\theta,\varphi) = f(\theta,\varphi) \quad \text{on the spherical surface } S,$$

using the Green's function mentioned above. To do so, we apply the Green's formula (2.183)(or its extension to a three-dimensional case), in which v is replaced by the Green's function (2.287), which yields

$$
u(r',\theta',\varphi') = \int_S\left(-u\frac{\partial G}{\partial n}\right)\mathrm{d}S = -\int_0^{\pi}\int_0^{2\pi} f(\theta,\varphi)\left(\frac{\partial G}{\partial r}\right)_{r=a} a^2\sin\theta\,\mathrm{d}\theta\,\mathrm{d}\varphi.
$$

A direct calculation reveals

$$
\left(\frac{\partial G}{\partial r}\right)_{r=a} = \frac{1}{4\pi a}\cdot\frac{r'^2 - a^2}{(a^2 - 2ar'\cos\omega + r'^2)^{3/2}},
$$

which are substituted for the previous equation to obtain the final result:

$$
u(r,\theta,\varphi) = \frac{a}{4\pi}\int_0^{\pi}\int_0^{2\pi} f(\theta',\varphi')\frac{a^2 - r^2}{(a^2 - 2ar\cos\omega + r^2)^{3/2}}\sin\theta'\mathrm{d}\theta'\mathrm{d}\varphi',
\tag{2.288}
$$

where the roles of (r',θ',φ') and (r,θ,φ) have been exchanged.

c. Dirichlet's problem on the wave motion in a spherical region

As an example, we consider a quantum mechanical particle confined in a sphere of radius a. The wave function that describes the particle motion is governed by the steady Schrödinger equation (see *e.g.*, [Schiff (1949); Arfken (1985); Beiser (1995)]:

$$
-\frac{\hbar^2}{2m}\triangle\psi = E\psi,
\tag{2.289}
$$

and the boundary conditions are

$$\psi(a) = 0, \qquad \psi : \text{finite inside the sphere } (r \le a),$$

where $\hbar$ is the Planck constant h divided by 2π $(i.e., \hbar = h/(2\pi))$, m is the mass, and E is the total energy.

Equation (2.289) is the Helmholtz-type ODE, and the part that depends on r is given by

$$\frac{d^2R}{dr^2} + \frac{2}{r}\frac{dR}{dr} + \left(k^2 - \frac{n(n+1)}{r^2}\right)R = 0, \tag{2.290}$$

where $k^2 = 2mE/\hbar^2$. The solution of the latter equation is given by the spherical Bessel functions (see Appendix A.2.3). Furthermore, if we focus our attention to the lowest energy of this particle, we may consider only the case of $n = 0$, so that the candidates of the solutions are $j_0(kr)$ and $n_0(kr)$. By the requirement of the finiteness of ψ inside the sphere, only the solution $j_0(kr)$ is allowable. From the boundary condition on the sphere, we have $ka = \alpha$, where α is the zeros of j_0. We know that the lowest zero of the spherical Bessel function j_0 is π, and hence $k = \pi/a$, so that we have

$$R(r) = Aj_0\left(\frac{\pi}{a}r\right), \qquad E = \frac{\hbar^2k^2}{2m} = \frac{h^2}{8ma^2}. \tag{2.291}$$

This result gives the lowest energy (zero-point energy) of the particle confined in a finite region, which is also the consequence of the uncertainty principle. (See *e.g.,* [Schiff (1949); Arfken (1985); Beiser (1995)] for further details.)

2.6.2 *Vibration of a circular membrane*

Let us consider the vibration of a circular membrane (radius a) whose periphery is fixed. The equation that governs the vibration is

$$\frac{1}{c^2}\frac{\partial^2 u}{\partial t^2} = \triangle u.$$

Taking account of the symmetry, it is appropriate to employ the polar coordinate system (r, θ) to describe the Laplacian:

$$\triangle u = \frac{\partial^2 u}{\partial r^2} + \frac{1}{r}\frac{\partial u}{\partial r} + \frac{1}{r^2}\frac{\partial^2 u}{\partial \theta^2}.$$

We assume the solution of the form $u(r, \theta) = \exp(i\omega t)\,R(r)\,\Theta(\theta)$, and separate the variables:

$$-\frac{1}{\Theta}\frac{d^2\Theta}{d\theta^2} = \frac{r^2}{R}\left(\frac{d^2R}{dr^2} + \frac{1}{r}\frac{dR}{dr}\right) + \frac{r^2}{R}\left(\frac{\omega}{c}\right)^2.$$

The l.h.s. depends only on θ, whereas r.h.s. depends only on r, so that we have

$$\frac{d^2\Theta}{d\theta^2} - \lambda\Theta = 0, \qquad \frac{d^2R}{dr^2} + \frac{1}{r}\frac{dR}{dr} + \left(\frac{\omega^2}{c^2} + \frac{\lambda}{r^2}\right)R = 0,$$

where λ is a constant of separation. From the physical requirement that $u(r, \theta)$ must be a smooth and one-valued periodic function of θ with period 2π, we have $\Theta = A_n \cos(n\theta) + B_n \sin(n\theta)$, where A_n and B_n are arbitrary constants and $\lambda = -n^2$ (n : integer). Consequently, the equation for R is

$$\frac{d^2 R}{dx^2} + \frac{1}{x}\frac{dR}{dx} + \left(1 - \frac{n^2}{x^2}\right) R = 0,$$

where $x = \omega r/c$. The latter is the Bessel's ODE. Among the solutions, the one that is finite at $x = 0$ is given by the Bessel function of the first kind $J_n(x)$ (see Eq. (A.11)), so that we obtain

$$u(r, \theta, t) = \exp(i\omega t)[A_n \cos(n\theta) + B_n \sin(n\theta)] J_n\left(\frac{\omega r}{c}\right).$$

The boundary condition requires that $\omega a/c$ is one of the zero points of $J_n(x)$. In terms of the m-th zero j_{nm} of the latter, the eigenfrequency of the vibration of the circular membrane is given by

$$\omega_{nm} = \frac{c}{a} j_{nm}.$$

In summary, the general solution of the wave equation for the membrane with a fixed boundary is

$$u(r, \theta, t) = \sum_{n=0}^{\infty} \sum_{m=1}^{\infty} J_n\left(j_{nm}\frac{r}{a}\right) \left\{ [A_{nm} \cos(n\theta) + B_{nm} \sin(n\theta)] \cos(\omega_{nm}t) \right.$$

$$\left. + [C_{nm} \cos(n\theta) + D_{nm} \sin(n\theta)] \sin(\omega_{nm}t) \right\}, \tag{2.292}$$

where the coefficients $A_{nm}, \ldots, D_{nm}$ are determined by the initial condition. Namely, if the latter is given as

$$u(r, \theta, 0) = f(r, \theta), \qquad \frac{\partial}{\partial t} u(r, \theta, 0) = g(r, \theta), \tag{2.293}$$

they are determined as (see Eqs. (A.22) and (A.23))

$$A_{nm} = \frac{2\varepsilon_n}{\pi a^2 J_{n+1}^2(j_{nm})} \int_0^a r J_n\left(j_{nm}\frac{r}{a}\right) dr \int_0^{2\pi} f(r, \theta) \cos(n\theta)\, d\theta,$$

$$B_{nm} = \frac{2}{\pi a^2 J_{n+1}^2(j_{nm})} \int_0^a r J_n\left(j_{nm}\frac{r}{a}\right) dr \int_0^{2\pi} f(r, \theta) \sin(n\theta)\, d\theta,$$

$$C_{nm} = \frac{1}{\omega_{mn}} \frac{2\varepsilon_n}{\pi a^2 J_{n+1}^2(j_{nm})} \int_0^a r J_n\left(j_{nm}\frac{r}{a}\right) dr \int_0^{2\pi} g(r, \theta) \cos(n\theta)\, d\theta,$$

$$D_{nm} = \frac{1}{\omega_{mn}} \frac{2}{\pi a^2 J_{n+1}^2(j_{nm})} \int_0^a r J_n\left(j_{nm}\frac{r}{a}\right) dr \int_0^{2\pi} g(r, \theta) \sin(n\theta)\, d\theta,$$

$$\tag{2.294}$$

where, $\varepsilon_n = 1/2$ ($n = 0$) and $\varepsilon_n = 1$ ($n \geq 1$). We show some examples of vibration modes in Fig. 2.18.

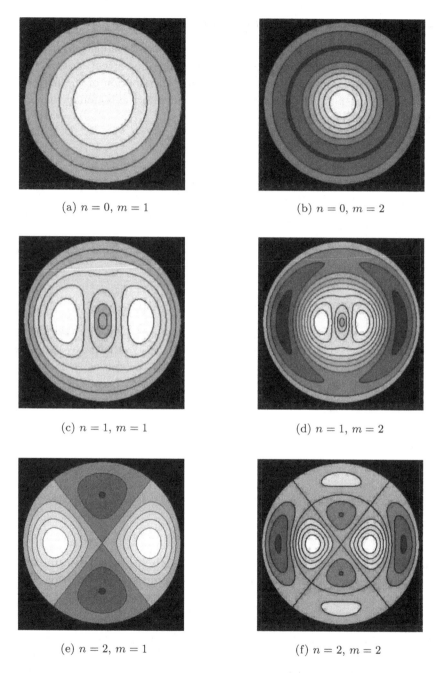

(a) $n = 0$, $m = 1$

(b) $n = 0$, $m = 2$

(c) $n = 1$, $m = 1$

(d) $n = 1$, $m = 2$

(e) $n = 2$, $m = 1$

(f) $n = 2$, $m = 2$

Fig. 2.18   Examples of vibration modes of the circular membrane.

2.6.3 *Diffusion in a circular region*

In this subsection, we consider the diffusion, or heat conduction, in a circular region. For simplicity, however, we confine our attention to an axisymmetric case, and assume that the density distribution of the material or temperature is finite within the region, and that the boundary value is kept constant. Then, the governing equation is

$$D\left(\frac{\partial^2 u}{\partial r^2} + \frac{1}{r}\frac{\partial u}{\partial r}\right) = \frac{\partial u}{\partial t}, \tag{2.295}$$

and the boundary conditions for this problem are

$$u = 0 \quad \text{at } r = a, \quad \text{and} \quad u \text{ is finite for } r < a;$$
$$u = f(r) \quad (\text{with } f(a) = 0 \text{ }) \quad \text{at } t = 0. \tag{2.296}$$

By normalizing r by a (*i.e.*, $x = r/a$), and assuming the solution of the form $u(x,t) = e^{-\lambda t}v(x)$, Eq. (2.295) becomes

$$\frac{d^2 v}{dx^2} + \frac{1}{x}\frac{dv}{dx} + k^2 v = 0, \quad \text{where} \quad k = \sqrt{\frac{\lambda a^2}{D}}. \tag{2.297}$$

The solution of this equation, which is finite at $x = 0$, is given by the Bessel function $J_0(kx)$. The boundary condition at $x = 1$ is satisfied by choosing $k = j_n$ ($n = 1, 2, \cdots$), where j_n is the n-th zero of J_0, so that we have

$$u(x,t) = \sum_{n=1}^{\infty} c_n J_0(j_n x) e^{-\lambda_n t}, \quad \text{where} \quad \lambda_n = \frac{D j_n^2}{a^2}. \tag{2.298}$$

The coefficient c_n is determined by the initial condition:

$$c_n = \frac{2}{J_1(j_n)^2} \int_0^1 x f(x) J_0(j_n x)\, dx. \tag{2.299}$$

See also the **Fourier–Bessel expansion** in §§2.3.4, or Appendix (A.23).

2.6.4 *Radiation and scattering of waves*

a. Solution of the Helmholtz equation with spherical symmetry
   Consider the spherically symmetric solution of the wave equation:

$$\Delta u = \frac{1}{c^2}\frac{\partial^2 u}{\partial t^2}. \tag{2.300}$$

By substitution of $u = \exp(-i\omega t)v(r)$ for Eq. (2.300), we have

$$(\Delta + k^2)v = 0, \quad \text{where} \quad \Delta = \frac{1}{r^2}\frac{d}{dr}\left(r^2\frac{d}{dr}\right), \quad k = \frac{\omega}{c}, \tag{2.301}$$

which is the Helmholtz-type ODE. If we notice the relation

$$\frac{d}{dr}\left(r^2\frac{dv}{dr}\right) = r\frac{d^2}{dr^2}(rv),$$

we can rewrite Eq. (2.301) as

$$\frac{d^2}{dr^2}(rv) + k^2(rv) = 0, \tag{2.302}$$

so that the solution $rv = \exp(\pm ikr)$ is easily obtained. The general solution of Eq. (2.301) is given by

$$v = A\frac{e^{ikr}}{r} + B\frac{e^{-ikr}}{r}, \tag{2.303}$$

and hence the general solution of Eq. (2.300) by

$$u = A\frac{e^{ik(r-ct)}}{r} + B\frac{e^{-ik(r+ct)}}{r}, \tag{2.304}$$

where A and B are arbitrary complex constants, and $\omega = kc$ is taken into account. Either term of Eq. (2.304) shows that a surface of equal phase is spherical. Therefore, if a certain spherical surface $r = a$ is the source of a wave at some instant, then the first term of the r.h.s. of Eq. (2.304) gives a wave with a spherical wave front propagating outwardly at a phase velocity c. This is a diverging spherical wave. Similarly, the second term of the r.h.s. gives a wave with a spherical wave front converging to the center with a phase velocity c.

When we focus to a wave propagating outwardly in an infinite region, e.g., an outgoing wave with initial condition $u = f(a)$, $\partial u/\partial t = g(a)$ on a spherical surface $r = a$ at time $t = 0$, we may use only the diverging wave given by the first term of the r.h.s. of Eq. (2.304). In terms of the real variables, the latter is given by

$$u = C_1\frac{\cos(kr - \omega t + \phi_1)}{r}, \tag{2.305}$$

where C_1 and ϕ_1 are real constants, which are determined by the initial and boundary conditions:

$$C_1 = af\sqrt{1 + \left(\frac{g}{\omega f}\right)^2}, \qquad \phi_1 = \arctan\left(\frac{g}{\omega f}\right) - ka. \tag{2.306}$$

Note that the respective functions of the r.h.s. of Eq. (2.303) are the Green's function of the Helmholtz ODE, *i.e.*,

$$(\triangle + k^2)\,G(\boldsymbol{x}, \boldsymbol{x}') = -\delta(\boldsymbol{x} - \boldsymbol{x}'), \tag{2.307}$$

where

$$G(\boldsymbol{x}, \boldsymbol{x}') = \frac{\exp(\pm\, \mathrm{i}k|\boldsymbol{x} - \boldsymbol{x}'|)}{4\pi|\boldsymbol{x} - \boldsymbol{x}'|}, \tag{2.308}$$

which are also derived later as an example of the Fourier transforms (§§3.3.2).

Making use of the Green's function above, we shall deal with the scattering of the incident waves later in this subsection **c**.

b. Solution of the axisymmetric Helmholtz equation

We consider the propagation of the axisymmetric waves. Similar to the previous subsection, we put $u = \exp(-\mathrm{i}\omega t)v(\rho)$ in the wave equation (2.300) in the cylindrical coordinate system (ρ, φ, z):

$$\frac{\mathrm{d}^2 v}{\mathrm{d}\rho^2} + \frac{1}{\rho}\frac{\mathrm{d}v}{\mathrm{d}\rho} + k^2 v = 0, \qquad k = \frac{\omega}{c}. \tag{2.309}$$

The solution is given by the 0-th order Bessel function (see Appendix, Eq. (A.10)). Therefore, the solution may be expressed as

$$v(\rho) = a J_0(k\rho) + b N_0(k\rho),$$

where a and b are real constants. However, taking account of the findings in a spherical wave where the asymptotic behavior is proportional to $\exp(\pm\mathrm{i}kr)$, we adopt the general solution of the type:

$$v(\rho) = A H_0^{(1)}(k\rho) + B H_0^{(2)}(k\rho), \tag{2.310}$$

where $H_0^{(1)}$ and $H_0^{(2)}$ are the Hankel functions (see Eq. (A.31)). As shown by Eq. (A.32) in Appendix A.2.2, their asymptotic behaviors are

$$\left.\begin{array}{c} H_0^{(1)}(k\rho) \\ H_0^{(2)}(k\rho) \end{array}\right\} \sim \sqrt{\frac{2}{\pi k\rho}}\, \exp\left[\pm\mathrm{i}\left(k\rho - \frac{\pi}{4}\right)\right], \tag{2.311}$$

so that $H_0^{(1)}$ describes the outgoing axisymmetric waves with phase velocity c, whereas $H_0^{(2)}$ describes the axisymmetric waves converging to the center. Note that $H_0^{(1)}$ and $H_0^{(2)}$ are, except for a proportional constant, the Green's functions of the Helmholtz ODE referred to in Table 2.2, column (b), two-dimensional case.

c. Scattering of the incident plane waves

It is well-known in quantum mechanics that the scattering of an incident plane wave $\Psi(\boldsymbol{x})_{\mathrm{in}} = \exp[\mathrm{i}(\boldsymbol{k} \cdot \boldsymbol{x} - \omega t)]$ by the steady state potential $V(\boldsymbol{x})$

is given by the following Schrödinger's equation (see *e.g.*, [Schiff (1949)], [Byron and Fuller (1969)](Chap. 7), [Beiser (1995)](Chap. 5)):

$$-\frac{\hbar^2}{2m} \triangle \Psi(\boldsymbol{x}) + V(\boldsymbol{x})\Psi(\boldsymbol{x}) = E\ \Psi(\boldsymbol{x}), \qquad (2.312)$$

which is rewritten as

$$(\triangle + k^2)\Psi(\boldsymbol{x}) = \frac{2m}{\hbar^2}V(\boldsymbol{x})\Psi(\boldsymbol{x}), \qquad (2.313)$$

where $k = \sqrt{2mE}/\hbar$.

We regard the r.h.s. as a perturbation, and expand Ψ as

$$\Psi = \Psi_{\text{in}} + \Psi_{\text{scat}}, \qquad (2.314)$$

where

$$\Psi_{\text{in}} \sim O(1), \qquad \Psi_{\text{scat}} \sim O(\epsilon), \qquad \text{with} \quad \epsilon = O\left(\frac{mV}{\hbar^2}\right).$$

By the expansion (2.314), Eq. (2.313) becomes

$$\begin{aligned} O(1): \quad & (\triangle + k^2)\Psi_{\text{in}}(\boldsymbol{x}) = 0, \\ O(\epsilon): \quad & (\triangle + k^2)\Psi_{\text{scat}}(\boldsymbol{x}) = \frac{2mV(\boldsymbol{x})}{\hbar^2}\Psi_{\text{in}}(\boldsymbol{x}). \end{aligned} \qquad (2.315)$$

The solution of the first equation is the incident plane wave $\exp[\,\mathrm{i}(\boldsymbol{k}\cdot\boldsymbol{x}-\omega t)]$ itself. If we substitute the latter for the r.h.s. of the second equation, we obtain the Helmholtz-type ODE with the known "external force" $(2mV/\hbar^2)\Psi_{\text{in}}$. To obtain the solution, we make use of the Green's function (2.308) of the outgoing-wave type, by which the scattered wave is given as

$$\Psi(\boldsymbol{x}) = \mathrm{e}^{\mathrm{i}(\boldsymbol{k}\cdot\boldsymbol{x}-\omega t)} - \frac{2m}{\hbar^2} \iiint \frac{\mathrm{e}^{\mathrm{i}k|\boldsymbol{x}-\boldsymbol{x}'|}}{4\pi|\boldsymbol{x}-\boldsymbol{x}'|}V(\boldsymbol{x}')\mathrm{e}^{\mathrm{i}(\boldsymbol{k}\cdot\boldsymbol{x}'-\omega t)}\mathrm{d}\boldsymbol{x}'.$$

A number of problems and solutions related to the present chapter can be found in *e.g.* [DuChateau and Zachmann (1986)].

Chapter 3

Integral Transforms and their Applications

In this chapter, we define and characterize the typical integral transforms that are useful for solving the partial differential equations (PDEs). In many cases, solutions in a transformed space are more easily obtained. Then, by applying the inverse transform to the latter, the desired solution in the original space is obtained. Thus far, many integral transforms are known. Readers will learn which integral transforms are appropriate under the given equations and boundary conditions.

3.1 General Theory of Integral Transforms and their Applicability

An integral $F(s)$:

$$F(s) \equiv \int_a^b f(x)\, K(s,x)\, \mathrm{d}x = \mathcal{I}\{f(x)\} \tag{3.1}$$

is called the **integral transform** of $f(x)$ with **kernel** $K(s,x)$. Various integral transforms are developed depending on (i) the types of the equation, and hence the types of kernel $K(s,x)$, (ii) intervals of the integral, and (iii) the functions to be transformed, among other factors.

Here, we confine our attention to the second-order PDEs (2.9) of §§2.2.1:

$$A\frac{\partial^2 u}{\partial x^2} + 2B\frac{\partial^2 u}{\partial x \partial y} + C\frac{\partial^2 u}{\partial y^2} + D\frac{\partial u}{\partial x} + E\frac{\partial u}{\partial y} + Fu = g. \tag{3.2}$$

Since Eq. (3.2) is symmetric with respect to variables x and y, we choose one of the variables x to exemplify the integral transform:

$$\mathcal{I}\{u(x,y)\} = \int_a^b u(x,y)\, K(s,x)\, \mathrm{d}x \equiv U(s,y). \tag{3.3}$$

We shall find the condition under which Eq. (3.2) is transformed to a *simpler* equation, *e.g.*, in the present case of PDE with two independent variables, transformed to an ODE of U which depends only on y. For simplicity, we first assume that $A, B, \cdots, F$ are allowed to be functions of y, but do not depend on x. By applying the integral transform (3.3) to Eq. (3.2), and taking account of the following calculations:

$$\mathcal{I}\left\{A(y)\frac{\partial^2 u}{\partial x^2}\right\} = A(y)\left(\left[\frac{\partial u}{\partial x}K - u\frac{\mathrm{d}K}{\mathrm{d}x}\right]_a^b + \int_a^b u\frac{\mathrm{d}^2 K}{\mathrm{d}x^2}\,\mathrm{d}x\right),$$

$$\mathcal{I}\left\{2B(y)\frac{\partial^2 u}{\partial x\partial y}\right\} = 2B(y)\left(\left[\frac{\partial u}{\partial y}K\right]_a^b - \int_a^b \frac{\partial u}{\partial y}\frac{\mathrm{d}K}{\mathrm{d}x}\,\mathrm{d}x\right),$$

$$\mathcal{I}\left\{C(y)\frac{\partial^2 u}{\partial y^2}\right\} = C(y)\frac{\mathrm{d}^2}{\mathrm{d}y^2}\int_a^b uK\,\mathrm{d}x = C(y)\frac{\mathrm{d}^2 U(s,y)}{\mathrm{d}y^2},$$

$$\mathcal{I}\left\{D(y)\frac{\partial u}{\partial x}\right\} = D(y)\left([uK]_a^b - \int_a^b u\frac{\mathrm{d}K}{\mathrm{d}x}\,\mathrm{d}x\right),\ \mathcal{I}\left\{E(y)\frac{\partial u}{\partial y}\right\} = E(y)\frac{\mathrm{d}U(s,y)}{\mathrm{d}y},$$

$$\mathcal{I}\{F(y)u\} = F(y)\int_a^b uK\,\mathrm{d}x = F(y)\,U(s,y), \quad \mathcal{I}\{g(x,y)\} = G(s,y),$$

we have

$$\left[A(y)\left(\frac{\partial u}{\partial x}K - u\frac{\mathrm{d}K}{\mathrm{d}x}\right) + 2B(y)\frac{\partial u}{\partial y}K + D(y)uK\right]_a^b$$
$$+ \int_a^b \left(A(y)u\frac{\mathrm{d}^2 K}{\mathrm{d}x^2} - 2B(y)\frac{\partial u}{\partial y}\frac{\mathrm{d}K}{\mathrm{d}x} - D(y)u\frac{\mathrm{d}K}{\mathrm{d}x}\right)\mathrm{d}x$$
$$+ C(y)\frac{\mathrm{d}^2 U(s,y)}{\mathrm{d}y^2} + E(y)\frac{\mathrm{d}U(s,y)}{\mathrm{d}y} + F(y)\,U(s,y) = G(s,y).$$

Here, we assume the convergence (existence) of integrals. Therefore, the sufficient conditions in which a PDE (3.2) is transformed to an ODE of U are

$$A(y)u\frac{\mathrm{d}^2 K}{\mathrm{d}x^2} - 2B(y)\frac{\partial u}{\partial y}\frac{\mathrm{d}K}{\mathrm{d}x} - D(y)u\frac{\mathrm{d}K}{\mathrm{d}x} \propto uK, \tag{3.4}$$

and

$$\left[A(y)\left(\frac{\partial u}{\partial x}K - u\frac{\mathrm{d}K}{\mathrm{d}x}\right) + 2B(y)\frac{\partial u}{\partial y}K + D(y)uK\right]_a^b = \text{known function of } y. \tag{3.5}$$

The division of Eq. (3.4) by u yields

$$A(y)\frac{\mathrm{d}^2 K}{\mathrm{d}x^2} - 2B(y)\frac{1}{u}\frac{\partial u}{\partial y}\frac{\mathrm{d}K}{\mathrm{d}x} - D(y)\frac{\mathrm{d}K}{\mathrm{d}x} \propto K, \tag{3.6}$$

which is satisfied only if $B = 0$ and A, D are independent of y (and hence the latter two are constant), because K is a function of x whereas u is a function of x and y. Then, Eqs. (3.4) and (3.5) become

$$A\frac{\mathrm{d}^2 K}{\mathrm{d}x^2} - D\frac{\mathrm{d}K}{\mathrm{d}x} = \lambda K, \tag{3.7}$$

$$\left[AK\frac{\partial u}{\partial x} + \left(DK - A\frac{\mathrm{d}K}{\mathrm{d}x} \right) u \right]_a^b = \text{known function of } y, \tag{3.8}$$

where λ is an arbitrary constant. (Other coefficients C, E, F are allowed to be functions of y as before.) ◁

We now relax the above constraints, and examine the case in which A, D are functions of x. By a similar calculation under these conditions, we have in place of Eqs. (3.7) and (3.8),

$$\frac{\mathrm{d}^2}{\mathrm{d}x^2}\left(A(x)K\right) - \frac{\mathrm{d}}{\mathrm{d}x}\left(D(x)K\right) = \lambda K, \tag{3.9}$$

$$\left[A(x)K\frac{\partial u}{\partial x} + \left(D(x)K - \frac{\mathrm{d}}{\mathrm{d}x}(A(x)K) \right) u \right]_a^b = \text{known function of } y. \tag{3.10}$$

Equation (3.9) is an ODE that determines the kernel K. Using thus determined K, and the boundary conditions (or initial conditions if variable x refers to time) on u and $\partial u/\partial x$, we may check whether or not Eq. (3.10) holds.

In summary, the conditions under which the integral transforms with respect to x are successfully applied to Eq. (3.2) are as follows:

(1) the equations are of the following type:

$$A(x)\frac{\partial^2 u}{\partial x^2} + C(y)\frac{\partial^2 u}{\partial y^2} + D(x)\frac{\partial u}{\partial x} + E(y)\frac{\partial u}{\partial y} + F(y)u = g(x, y),$$

(2) the kernel satisfies Eq. (3.9),
(3) the boundary (or initial) conditions satisfy Eq. (3.10). ◁

Example 3.1. When A and D are constants, Eq. (3.9) becomes

$$A\frac{\mathrm{d}^2 K}{\mathrm{d}x^2} - D\frac{\mathrm{d}K}{\mathrm{d}x} = \lambda K.$$

This is a homogeneous ODE with constant coefficients, whose solution is of the form $K \propto \exp(\alpha x)$. As is shown in a later section (§3.3), the choice of $K = \exp(ikx)$ $(-\infty < x < \infty)$ provides the Fourier transform, whereas the choice of $K = \exp(-sx)$ $(0 \leq x < \infty)$ provides the Laplace transform. ◁

Example 3.2. When $A = 1$, $D = -1/x$ and $\lambda = -k^2$, Eq. (3.9) becomes

$$\frac{\mathrm{d}^2 K}{\mathrm{d}x^2} + \frac{1}{x}\frac{\mathrm{d}K}{\mathrm{d}x} + \left(k^2 - \frac{1}{x^2}\right)K = 0.$$

The solution of the above equation is given by the Bessel function $J_1(kx)$, and the use of this kernel leads to the Hankel transform. ◁

These examples show that the choice of kernels and intervals, and hence the choice of the integral transforms used to solve the given PDEs, depends on the type of equation, geometry of the region considered, and the conditions imposed at the boundary, among other factors, which we examine in detail in the following sections. ◁

Some general remarks:

First, the integral transforms thus far referred to are linear, *i.e.*, they satisfy

$$\int_a^b cf(x)\,K(s,x)\,\mathrm{d}x = c\int_a^b f(x)\,K(s,x)\,\mathrm{d}x, \tag{3.11}$$

$$\int_a^b [f_1(x) + f_2(x)]\,K(s,x)\,\mathrm{d}x = \int_a^b f_1(x)\,K(s,x)\,\mathrm{d}x + \int_a^b f_2(x)\,K(s,x)\,\mathrm{d}x, \tag{3.12}$$

where c is a constant, and f, f_1, f_2 are arbitrary functions.

Second, if Eq. (3.1):

$$F(s) = \mathcal{I}\{f(x)\}, \tag{3.13}$$

is defined, where $\mathcal{I}$ is a linear operator relevant to the given equation, we may expect the existence of its inverse transform $\mathcal{I}^{-1}$, so that the solution is given by

$$f(x) = \mathcal{I}^{-1}\{F(s)\}. \tag{3.14}$$

The proof on the existence of the inverse transform, however, is generally not easy. In this book, we do not go into detail in this area except when necessary in the course of our calculations.

We shall see in later sections that the solution procedures become easier in the transformed space even if the solution procedures are difficult in the original space. For instance, a given ODE may be reduced to an algebraic equation in a transformed space, or the given PDEs in the original space may be reduced to ODEs, *etc.* In these cases, we can obtain a solution relatively easily in a transformed space, which are inversely transformed to attain the solution in the original space (see Fig. 3.1).

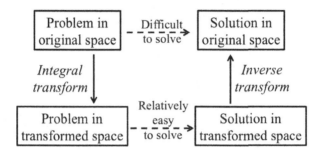

Fig. 3.1   Application of integral transforms.

3.2   Integral Transforms in a Finite Region

3.2.1   *General theory*

We examine here the integral transforms defined in the finite interval $0 \leq x \leq L$. We pay particular attention to the boundary conditions, and consider the choice of kernel.

As a simple example, if we put $A = 1$, $D = 0$, $\lambda = -\Lambda^2$ (constant), $a = 0$, $b = L$ in Eqs. (3.9) and (3.10), then they respectively become,

$$\frac{\mathrm{d}^2 K}{\mathrm{d}x^2} + \Lambda^2 K = 0, \tag{3.15}$$

$$\left[K \frac{\partial u}{\partial x} - \frac{\mathrm{d}K}{\mathrm{d}x} u \right]_0^L = \text{known function of } y. \tag{3.16}$$

Solutions of Eq. (3.15) are given by trigonometric functions. If we assume furthermore that the right-hand side of Eq. (3.16) is zero, then K and u, or $\mathrm{d}K/\mathrm{d}x$ and $\partial u/\partial x$ must vanish at $x = 0$ and L, *i.e.*,

$$K = \sin\left(\frac{n\pi x}{L}\right), \qquad u = 0 \quad \text{at} \quad x = 0, L, \tag{3.17}$$

or
$$K = \cos\left(\frac{n\pi x}{L}\right), \quad \frac{\partial u}{\partial x} = 0 \quad \text{at} \quad x = 0, \, L. \tag{3.18}$$

◁

Example 3.3. Solve the one-dimensional wave equation:
$$\frac{\partial^2 u}{\partial x^2} = \frac{1}{c^2}\frac{\partial^2 u}{\partial t^2}, \tag{3.19}$$

under the boundary conditions
$$u(0, t) = u(L, t) = 0, \tag{3.20}$$

and initial conditions
$$u(x, 0) = u_0(x), \quad \frac{\partial u}{\partial t}(x, 0) = v_0(x). \tag{3.21}$$

Taking account of the boundary conditions, we may expect a solution of the form
$$u(x, t) = \sum_{n=1}^{\infty} U_n \sin\left(\frac{n\pi x}{L}\right), \quad U_n = \frac{2}{L}\int_0^L u(x, t)\sin\left(\frac{n\pi x}{L}\right)dx,$$

where U_n is the same as Eq. (2.161) determined as an example of eigenfunction expansion (§§2.3.3). The latter relation is also regarded as an integral transform defined in the interval $[0, L]$ with kernel $K = \sin(n\pi x/L)$, so that we shall introduce the following finite integral transform:
$$\mathcal{L}\{u(x, t)\} = \int_0^L u(x, t)\sin\left(\frac{n\pi x}{L}\right)dx = U(n, t)\left(=\frac{L}{2}U_n\right). \tag{3.22}$$

By applying this integral transform to Eq. (3.19), and noting the following relations:
$$\mathcal{L}\left\{\frac{\partial^2}{\partial x^2}u(x, t)\right\} = -\left(\frac{n\pi}{L}\right)^2 U(n, t), \mathcal{L}\left\{\frac{\partial^2}{\partial t^2}u(x, t)\right\} = \frac{d^2}{dt^2}U(n, t), \tag{3.23}$$

we obtain
$$\frac{1}{c^2}\frac{d^2}{dt^2}U = -\left(\frac{n\pi}{L}\right)^2 U. \tag{3.24}$$

The solution of Eq. (3.24) is
$$U(n, t) = \tilde{A}_n \cos\left(\frac{n\pi ct}{L}\right) + \tilde{B}_n \sin\left(\frac{n\pi ct}{L}\right), \tag{3.25}$$

where $\tilde{A}_n$ and $\tilde{B}_n$ are arbitrary constants. Consequently, we obtain the solution:
$$u(x, t) = \sum_{n=1}^{\infty}\left[A_n \cos\left(\frac{n\pi ct}{L}\right) + B_n \sin\left(\frac{n\pi ct}{L}\right)\right]\sin\left(\frac{n\pi x}{L}\right), \tag{3.26}$$

where A_n and B_n are constants to be determined by the initial conditions. The last expression is the same as Eq. (2.160).

◁

Although the method of solution by the use of a general finite integral transform is similarly developed, the latter is equivalent to the eigenfunction expansion, details of which are given in §§2.3.3.

3.2.2 *Further applications*

Let us consider the vibration of a rectangular membrane with sides a and b. The basic equation and boundary conditions are

$$\frac{1}{c^2}\frac{\partial^2 u}{\partial t^2} = \frac{\partial^2 u}{\partial x^2} + \frac{\partial^2 u}{\partial y^2}, \tag{3.27}$$

$$u(0,y,t) = u(a,y,t) = 0, \qquad u(x,0,t) = u(x,b,t) = 0. \tag{3.28}$$

Taking account of the boundary conditions, we consider the integral transform

$$\mathcal{L}\{u(x,y,t)\} = \int_0^b \int_0^a u(x,y,t)\sin\left(\frac{m\pi x}{a}\right)\sin\left(\frac{n\pi y}{b}\right)\mathrm{d}x\,\mathrm{d}y$$

$$= U(m,n,t). \tag{3.29}$$

By applying the above integral transform to Eq. (3.27), we have

$$\frac{1}{c^2}\frac{\mathrm{d}^2 U}{\mathrm{d}t^2} = -\left[\left(\frac{m\pi}{a}\right)^2 + \left(\frac{n\pi}{b}\right)^2\right]U, \tag{3.30}$$

from which we obtain the fundamental solution:

$$U(m,n,t) = A_{mn}\cos(\omega_{mn}t + \phi_{mn}), \qquad \omega_{mn} = \pi c\sqrt{\left(\frac{m}{a}\right)^2 + \left(\frac{n}{b}\right)^2}, \tag{3.31}$$

and the general solution:

$$u(x,y,t) = \sum_{m=1}^{\infty}\sum_{n=1}^{\infty} A_{mn}\sin\left(\frac{m\pi x}{a}\right)\sin\left(\frac{n\pi y}{b}\right)\cos(\omega_{mn}t + \phi_{mn}). \tag{3.32}$$

Here, the amplitude A_{mn} and initial phase ϕ_{mn} are constants to be determined by the initial conditions. If the latter conditions are given as

$$u(x,y,0) = u_0(x,y), \qquad \frac{\partial}{\partial t}u(x,y,0) = v_0(x,y), \tag{3.33}$$

they are determined as

$$A_{mn}\cos\phi_{mn} = \frac{4}{ab}\int_0^a\int_0^b u_0(x,y)\sin\left(\frac{m\pi x}{a}\right)\sin\left(\frac{n\pi y}{b}\right)\mathrm{d}x\,\mathrm{d}y, \tag{3.34}$$

$$A_{mn}\sin\phi_{mn} = -\frac{4}{ab\,\omega_{mn}}\int_0^a\int_0^b v_0(x,y)\sin\left(\frac{m\pi x}{a}\right)\sin\left(\frac{n\pi y}{b}\right)\mathrm{d}x\,\mathrm{d}y. \tag{3.35}$$

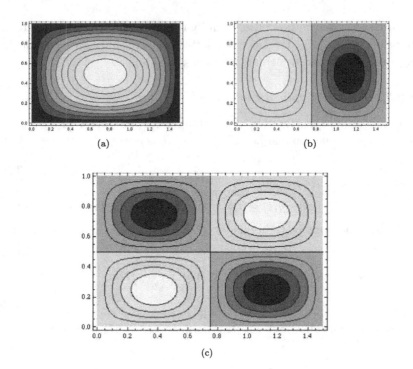

Fig. 3.2 Examples of eigenmodes in a rectangular membrane ($a = 1.5$, $b = 1$). (a) $m = n = 1$, (b) $m = 2, n = 1$, and (c) $m = n = 2$.

Here, $\omega_{mn}/(2\pi)$ is the eigenfrequency. We show some examples of an eigenmode in Fig. 3.2.   ◁

In a rectangular membrane, different vibration modes specified by the set (m, n) with the same $(m/a)^2 + (n/b)^2$ value can coexist. The states that have the same frequency but are of different vibration modes are called **degenerate** to each other. We show examples in the square membrane $(a = b)$ with $m = 2$, $n = 1$ and $m = 1$, $n = 2$ in Fig. 3.3.

3.3 Integral Transforms in an Infinite Region

3.3.1 *General theory*

In general, the integral

$$F(s) = \int_{-\infty}^{\infty} f(x) K(s, x) \, \mathrm{d}x \tag{3.36}$$

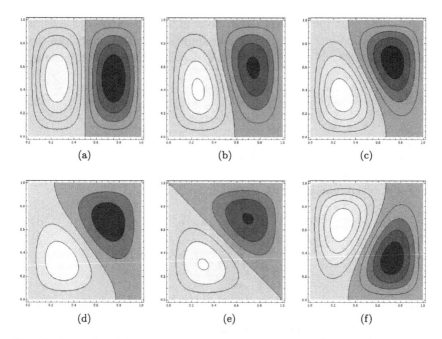

(a)   (b)   (c)

(d)   (e)   (f)

Fig. 3.3 Examples of degenerate states in a square membrane ($a = b = 1$) corresponding to $m = 2$, $n = 1$ and $m = 1$, $n = 2$ with a different ratio $\beta = A_{21}/A_{12} [i.e.,$ $(2,1)$ mode$/(1,2)$ mode]. The contour lines of $(2,1)$ mode $+ \beta \times (1,2)$ mode : (a) $\beta = 0$, (b) $\beta = 0.25$, (c) $\beta = 0.5$, (d) $\beta = 0.75$, (e) $\beta = 1$, and (f) $\beta = -0.5$. [Note that negative values of β provides a mirror symmetric pattern with respect to $y = 1/2$, as in (c) versus (f).]

is called an **infinite integral transform** of $f(x)$ with **kernel** $K(s, x)$. In the following, we show some examples that are frequently used.

a. Fourier transforms and their inverse transforms

The integral

$$\mathcal{F}\{u(x)\} \equiv \frac{1}{\sqrt{2\pi}} \int_{-\infty}^{\infty} u(x)\, e^{-ikx} dx = U(k) \tag{3.37}$$

is called the **Fourier transform** of $u(x)$, and

$$u(x) = \frac{1}{\sqrt{2\pi}} \int_{-\infty}^{\infty} U(k)\, e^{ikx} dk = \mathcal{F}^{-1}\{U(k)\} \tag{3.38}$$

is called the **inverse Fourier transform**. Here, we assume that $u(x)$ is a piecewise continuous function in the range $-\infty < x < \infty$, is a function of bounded variation at an arbitrary finite interval, and the integral

$$\int_{-\infty}^{\infty} |u(x)|\, dx$$

exists. Note that the integral (3.38) gives $(1/2)[u(x-0) + u(x+0)]$ at a discontinuous point x (a similar remark applies hereafter). Note also that a set of Eqs. (3.37) and (3.38) is a two-step description of the Fourier integral formula

$$u(x) = \frac{1}{2\pi} \int_{-\infty}^{\infty} dk \int_{-\infty}^{\infty} u(\xi)\, e^{ik(x-\xi)}\, d\xi. \tag{3.39}$$

◁

Viewing the reciprocal relations between Eq. (3.37) and Eq. (3.38), we may call $u(x)$ the Fourier transform of $U(k)$, and *vice versa*.

For an even function $u(x)$, *i.e.*, $u(-x) = u(x)$,

$$\frac{1}{\sqrt{2\pi}} \int_{-\infty}^{\infty} u(x)\, e^{-ikx} dx = \frac{1}{\sqrt{2\pi}} \left(\int_{-\infty}^{0} u(x)\, e^{-ikx} dx + \int_{0}^{\infty} u(x)\, e^{-ikx} dx \right)$$

$$= \frac{1}{\sqrt{2\pi}} \int_{0}^{\infty} u(x)(e^{ikx} + e^{-ikx})\, dx$$

$$= \sqrt{\frac{2}{\pi}} \int_{0}^{\infty} u(x)\, \cos(kx)\, dx,$$

so that Eq. (3.37) may be rewritten as

$$U_c(k) = \sqrt{\frac{2}{\pi}} \int_{0}^{\infty} u(x)\, \cos(kx)\, dx \equiv \mathcal{F}_c\{u(x)\}, \tag{3.40}$$

which is called the **Fourier cosine transform**. Note that the subscript "c" reflects the difference in the kernel function and the range of integral. The inverse Fourier cosine transform is given as

$$u(x) = \sqrt{\frac{2}{\pi}} \int_{0}^{\infty} U_c(k)\, \cos(kx)\, dk. \tag{3.41}$$

For an odd function $u(x)$, *i.e.*, $u(-x) = -u(x)$, we can similarly define the **Fourier sine transform**

$$U_s(k) \equiv \sqrt{\frac{2}{\pi}} \int_{0}^{\infty} u(x)\, \sin(kx)\, dx \equiv \mathcal{F}_s\{u(x)\}, \tag{3.42}$$

and its inverse transform

$$u(x) = \sqrt{\frac{2}{\pi}} \int_{0}^{\infty} U_s(k)\, \sin(kx)\, dk. \tag{3.43}$$

Note that the Fourier integral formula in terms of trigonometric functions is given by

$$u(x) = \frac{1}{\pi} \int_{0}^{\infty} dk \int_{-\infty}^{\infty} u(\xi)\, \cos[k(x-\xi)]\, d\xi. \tag{3.44}$$

b. Laplace transforms and their inverse transforms

The integral

$$\mathcal{L}\{u(t)\} \equiv \int_0^\infty u(t)\,\mathrm{e}^{-st}\mathrm{d}t = U(s) \tag{3.45}$$

is called the **Laplace transform** of $u(t)$, and

$$u(t) = \frac{1}{2\pi i}\int_{\gamma-i\infty}^{\gamma+i\infty} U(s)\,\mathrm{e}^{st}\,\mathrm{d}s = \mathcal{L}^{-1}\{U(s)\} \tag{3.46}$$

is called the **inverse Laplace transform** of $U(s)$.[1]

This integral transform is applied to a function u, which is piecewise continuous for $t > 0$, and is of bounded variation in an arbitrary finite interval. The existence of a constant a such that $u(t)\mathrm{e}^{-at}$ is bounded for $t \to \infty$ is assumed. Under these conditions, the inverse Laplace transform is given by the contour integral in the complex-s plane with $\gamma > a$. See the **Bromwich integral** in Eq. (3.139).

The set of Eqs. (3.45) and (3.46) is a two-step description of the Laplace–Mellin integral formula

$$u(t) = \frac{1}{2\pi i}\int_{\gamma-i\infty}^{\gamma+i\infty} \mathrm{d}s\,\mathrm{e}^{st}\int_0^\infty u(\tau)\,\mathrm{e}^{-s\tau}\mathrm{d}\tau. \tag{3.47}$$

c. Mellin transforms and their inverse transforms

The integral

$$\mathcal{M}\{u(x)\} \equiv \int_0^\infty u(x)\,x^{s-1}\mathrm{d}x = U(s) \tag{3.48}$$

is called the **Mellin transform** of $u(t)$, and

$$u(x) = \frac{1}{2\pi i}\int_{\gamma-i\infty}^{\gamma+i\infty} U(s)\,x^{-s}\mathrm{d}s = \mathcal{M}^{-1}\{U(s)\} \quad (0 < \gamma < 1) \tag{3.49}$$

is called the **inverse Mellin transform** of $U(s)$. The set of Eqs. (3.48) and (3.49) is a two-step description of the Mellin integral formula

$$u(x) = \frac{1}{2\pi i}\int_{\gamma-i\infty}^{\gamma+i\infty} \mathrm{d}s\,x^{-s}\int_0^\infty u(\xi)\,\xi^{s-1}\mathrm{d}\xi \quad (0 < \gamma < 1). \tag{3.50}$$

---

[1]A slightly different definition is sometimes used. In the latter, the Laplace transform of $f(t)$ is defined as

$$\mathcal{L}\{f(t)\} = p\int_0^\infty f(t)\mathrm{e}^{-pt}\mathrm{d}t = F(p),$$

which is termed the "p-multiplied Laplace transform" (see *e.g.*, [McLachlan (1962)]).

d. Hankel transforms and their inverse transforms

The integral

$$\mathcal{H}\{u(x)\} \equiv \int_0^\infty u(x)\, x\, J_n(\xi x)\ \mathrm{d}x = U(\xi) \tag{3.51}$$

is called the **Hankel transform** of $u(x)$, and

$$u(x) = \int_0^\infty U(\xi)\, \xi\, J_n(x\xi)\ \mathrm{d}\xi = \mathcal{H}^{-1}\{U(\xi)\} \tag{3.52}$$

is called the **inverse Hankel transform** of $U(\xi)$, where $J_n(x)$ is the n-th order Bessel function of the first kind. The set of Eqs. (3.51) and (3.52) is a two-step description of the Fourier–Bessel integral formula

$$u(x) = \int_0^\infty \mathrm{d}\xi\ \xi\, J_n(\xi x) \int_0^\infty u(\eta)\, \eta\, J_n(\xi\eta)\ \mathrm{d}\eta. \tag{3.53}$$

All of these integral transforms are linear. In the following, we consider each transform in further detail. Which integral transform is appropriate depends on the phenomena in question, as well as on the conditions on the boundary. Roughly speaking, the problems in which the boundary conditions are specified on a plane or a straight line in an infinite region, or the problems in which the phenomena continue indefinitely in time, are more often analyzed by the Fourier transforms. Meanwhile, the problems in which the boundary condition is specified in a semi-infinite region, or initial-value problems, are appropriately analyzed by the Laplace transforms; the boundary-value problems in which a wedge-shaped region extends to infinity are suitably analyzed by the Mellin transforms; and the boundary value problems in which conditions are specified on a circle or a circular cylindrical surface may well be analyzed by the Hankel transforms. For a more rigorous mathematical background, see *e.g.*, [Titchmarsh (1948)]. In addition to the integral transforms mentioned above, there are other types of integral transforms that may be appropriate for respective purposes (see *e.g.*, [Erdélyi (1954)]).

3.3.2 *Fourier transforms*

In §§2.1.1 we derived a one-dimensional wave equation

$$\frac{\partial^2 u}{\partial t^2} = c^2 \frac{\partial^2 u}{\partial x^2}. \tag{3.54}$$

We learned thus far that a wave with wavenumber k propagating in the x direction is described by a fundamental wave of the form such as

$\sin(kx)$, $\cos(kx)$, and $\exp(\pm ikx)$. We also learned that any given waves are described by a superposition of the set of fundamental waves with suitably determined amplitudes.

When we deal with the waves in a finite interval, only the discrete wavenumber k_n that satisfies the boundary condition is allowed, so that the wave in question is described by a summation of the set of waves with amplitudes specified to each wave, leading to the Fourier series. However, when we consider the waves in an infinite region, the wavenumber becomes continuous, so that the summation is replaced by the integral with respect to wavenumber k. Taking account of the dependence on time t of the wave, the waves of an infinite extension are described as

$$u(x,t) \propto \int_{-\infty}^{\infty} U(k,t) \left\{ \begin{array}{c} \sin(kx) \\ \cos(kx) \\ e^{\pm ikx} \end{array} \right\} dk,$$

where $U(k,t)$ is the amplitude of the fundamental waves. Except for the proportional constant, this expression is the Fourier transform itself. As the Fourier transforms have been dealt with in quite a number of textbooks, we show here only the basic properties.

a. Simple examples
(i) Fourier cosine transforms
   We show some simple examples obtained by Eq. (3.40).
(1) Exponential function

$$\mathcal{F}_c\left\{e^{-ax}\right\} \equiv \sqrt{\frac{2}{\pi}} \int_0^{\infty} e^{-ax} \cos(kx)\, dx = \sqrt{\frac{2}{\pi}} \frac{a}{k^2 + a^2}, \quad (a > 0).$$

(2) Gaussian-type function

$$\mathcal{F}_c\left\{e^{-ax^2}\right\} = \frac{1}{\sqrt{2a}} \exp\left(-\frac{k^2}{4a}\right), \quad (a > 0),$$

in particular

$$\mathcal{F}_c\left\{\exp\left(-\frac{x^2}{2}\right)\right\} = \exp\left(-\frac{k^2}{2}\right).$$

(3) Function of the powers of x

$$\mathcal{F}_c\{x^{a-1}\} = \sqrt{\frac{2}{\pi}} \Gamma(a) \cos\left(\frac{\pi a}{2}\right) k^{-a}, \quad (0 < a < 1),$$

where $\Gamma(x)$ is the **gamma function** (see Appendix A.1.1). To obtain the last relation, we perform a contour integral of the complex function

$z^{a-1}\exp(-kz)$. Here, taking into account that $z=0$ is a branch point, we choose the closed contour C consisting of the following four parts: C_ϵ (a quarter circle of infinitesimal radius ϵ in the first quadrant that surrounds the branch point, *i.e.*, $z=\epsilon e^{i\theta}, \theta=\pi/2\to0$), C_1 (a line segment on the real axis x, such that $x=\epsilon\to R$ ($R\gg1$)), C_R (a quarter circle of radius R, such that $z=Re^{i\theta}, \theta=0\to\pi/2$), and C_2 (a line segment on the imaginary axis, such that $z=ir, r=R\to\epsilon$). Since there is no singular point inside the closed contour,

$$\int_C z^{a-1}e^{-kz}dz=0.$$

Meanwhile, the integral on the contour C_ϵ and that on C_R vanish as $\epsilon\to0$ and $R\to\infty$, respectively, for $0<a<1$. The contributions on the contour C_1 and C_2 are

$$\int_0^\infty x^{a-1}e^{-kx}\,dx-\exp\left(\frac{\pi}{2}ia\right)\int_0^\infty r^{a-1}e^{-ikr}dr=0,$$

as $\epsilon\to0$ and $R\to\infty$. The first term of the l.h.s. of the above equation is equal to $k^{-a}\Gamma(a)$ (see Eq. (A.1)). Hence, we have

$$\int_0^\infty r^{a-1}e^{-ikr}dr=\exp\left(-\frac{\pi}{2}ia\right)k^{-a}\Gamma(a),$$

and the real part of which yields[2]

$$\int_0^\infty x^{a-1}\cos(kx)dx=k^{-a}\Gamma(a)\cos\left(\frac{\pi a}{2}\right).$$

Note that we have shown the result here as an example of the Fourier cosine transform of x^{a-1}; however, the same holds as an example of the Mellin transform of $\cos(kx)$ (see §§3.3.4 **a** (vii)).

(4) Unit-step-type function

The function defined by

$$f(x)=\begin{cases}1 & (0\le x<a)\\0 & (x>a)\end{cases}=\mathbf{1}(x)-\mathbf{1}(x-a),$$

where $\mathbf{1}(x)$ is the unit step function, is Fourier cosine transformed as

$$\mathcal{F}_c\{f(x)\}=\sqrt{\frac{2}{\pi}}\int_0^a\cos(kx)\,dx=\sqrt{\frac{2}{\pi}}\frac{\sin(ka)}{k}.$$

---

[2]The imaginary part of this equation yields

$$\int_0^\infty x^{a-1}\sin(kx)dx=k^{-a}\Gamma(a)\sin\left(\frac{\pi a}{2}\right),$$

which provides an example of the Fourier sine transform of x^{a-1}.

(5) Bessel function J_0 (0-th order, first kind)

$$\mathcal{F}_c\left\{J_0(ax)\right\} = \sqrt{\frac{2}{\pi}}\,\frac{1}{\sqrt{a^2 - k^2}}\,\mathbf{1}(a - k), \quad (a > k > 0).$$

The derivation of $\mathcal{F}_c\left\{J_0(x)\right\}$ is given in a later subsection dealing with the Hankel transform (see §§3.3.5a, Eq. (3.183)).

(ii) Fourier sine transforms
    We show some simple examples obtained by Eq. (3.42).
(1) Exponential function

$$\mathcal{F}_s\left\{e^{-ax}\right\} \equiv \sqrt{\frac{2}{\pi}}\int_0^\infty e^{-ax}\sin(kx)\,\mathrm{d}x = \sqrt{\frac{2}{\pi}}\,\frac{k}{k^2 + a^2}, \quad (a > 0).$$

(2) Gaussian-type function

$$\mathcal{F}_s\left\{xe^{-ax^2}\right\} = \frac{k}{(2a)^{3/2}}\exp\left(-\frac{k^2}{4a}\right), \quad (a > 0),$$

in particular

$$\mathcal{F}_s\left\{x\exp\left(-\frac{x^2}{2}\right)\right\} = k\exp\left(-\frac{k^2}{2}\right).$$

(3) Bessel function J_0

$$\mathcal{F}_s\left\{J_0(ax)\right\} = \sqrt{\frac{2}{\pi}}\,\frac{1}{\sqrt{k^2 - a^2}}\,\mathbf{1}(k - a), \quad (k > a > 0).$$

The derivation of $\mathcal{F}_s\left\{J_0(x)\right\}$ is given in a later subsection dealing with the Hankel transform (see §§3.3.5a, Eq. (3.184)).
(4) On integrating by parts of the previous result §§3.3.2a(i)-(5), we have

$$\mathcal{F}_s\left\{J_1(ax)\right\} = \sqrt{\frac{2}{\pi}}\int_0^\infty J_1(ax)\sin(kx)\,\mathrm{d}x = \sqrt{\frac{2}{\pi}}\,\frac{k/a}{\sqrt{a^2 - k^2}}\,\mathbf{1}(a - k)$$

for $0 < k < a$. The latter is also an example of the Hankel transform (see Eq. (3.186)).

(iii) Fourier transforms
    The following are some examples of the Fourier transforms with a kernel of the type $e^{\pm ikx}$ defined by Eq. (3.37).
(1) Lorentzian-type function

$$\mathcal{F}\left\{\frac{a}{x^2 + a^2}\right\} = \sqrt{\frac{\pi}{2}}\,e^{-a|k|}, \quad (a > 0).$$

This example is the same as Eqs. (2.204) and (2.211) shown in §§2.4.4, except for the proportional constant.

(2) Gaussian-type function

$$\mathcal{F}\left\{e^{-ax^2}\right\} = \frac{1}{\sqrt{2a}}\exp\left(-\frac{k^2}{4a}\right).$$

This result is the same as that given in the Fourier cosine transform, because the integrand is an even function. We show the curves of the integrand and their transforms for several values of a in Fig. 3.4. Note that the steeper curve in one space corresponds to the broader curve in the other, which is the characteristics of the Fourier transforms.

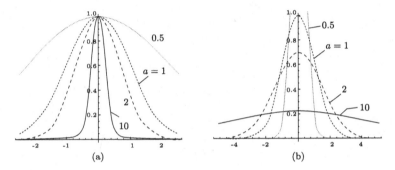

Fig. 3.4   (a) $\exp(-ax^2)$ and (b) its Fourier transform: $a = 0.5$ (thin dotted line), $a = 1$ (dotted line), $a = 2$ (broken line), and $a = 10$ (solid line).

(3) Delta function. 1

A delta function has already been defined by Eq. (2.176):

$$\delta(x) = \frac{1}{2\pi}\int_{-\infty}^{\infty} e^{ikx}\, dk,$$

and is given as a limit $n \to \infty$ of the function depicted in Fig. 2.9(d). The Fourier transform of this function is

$$\mathcal{F}\{\delta(x)\} = \frac{1}{\sqrt{2\pi}}\int_{-\infty}^{\infty}\delta(x)e^{-ikx}\, dx = \frac{1}{\sqrt{2\pi}},$$

so that $\delta(x)$ and $1/\sqrt{2\pi}$ are the Fourier transform and the inverse Fourier transform of the other.

(4) Delta function. 2

Using the definition of a delta function described by Fig. 2.9(a), as a limit $n \to \infty$ of

$$f(x) = \begin{cases} n & (|x| < 1/2n) \\ 0 & (|x| > 1/2n) \end{cases},$$

we obtain the Fourier transform:

$$\mathcal{F}\{f(x)\} = \frac{1}{\sqrt{2\pi}} \int_{-1/(2n)}^{1/(2n)} n e^{-ikx} \, dx = \frac{1}{\sqrt{2\pi}} \frac{2n}{k} \sin\left(\frac{k}{2n}\right),$$

which tends to $1/\sqrt{2\pi}$ as $n \to \infty$, in agreement with the transform of the delta function mentioned above.

(5) Delta function. 3

Consider the definition of a delta function given by Fig. 2.9(b), *i.e.*, the one given as a limit $n \to \infty$ of the function series $\delta_n(x) = n/\sqrt{\pi} \exp\left(-n^2 x^2\right)$. The Fourier transform of $\delta_n(x)$ is

$$\mathcal{F}\{\delta_n(x)\} = \frac{n}{\sqrt{2}\,\pi} \int_{-\infty}^{\infty} \exp(-n^2 x^2 - ikx) \, dx = \frac{1}{\sqrt{2\pi}} \exp\left(-\frac{k^2}{4n^2}\right),$$

which tends to $1/\sqrt{2\pi}$ as $n \to \infty$, in agreement with the transform of the delta function mentioned above.

(6) Delta function. 4

Consider the definition of a delta function given by Fig. 2.9(c). Here, the delta function is given as a limit $n \to \infty$ of the function series $\delta_n(x) = n/\pi(1 + n^2 x^2)$. The Fourier transform of the latter is

$$\mathcal{F}\{\delta_n(x)\} = \frac{1}{\sqrt{2\pi}} \int_{-\infty}^{\infty} \frac{n}{\pi} \frac{1}{1 + n^2 x^2} \exp(-ikx) \, dx = \frac{1}{\sqrt{2\pi}} \exp\left(-\frac{|k|}{n}\right),$$

which tends to $1/\sqrt{2\pi}$ as $n \to \infty$, in agreement with the transform of the delta function mentioned above.

(7) Miscellaneous

The function $f(x)$ defined by

$$f(x) = \begin{cases} \sqrt{\dfrac{2}{\pi}} \dfrac{1}{\sqrt{a^2 - x^2}} & (|x| < a) \\ 0 & (|x| > a) \end{cases}$$

is Fourier transformed to

$$\mathcal{F}\{f(x)\} = J_0(ka),$$

which is the same as example §§3.3.2a(i)-(5) of the Fourier cosine transform, as should be for an even function. For an alternative derivation of this result, we may transform the variable x to θ such as $x = a \sin\theta$

$$\mathcal{F}\{f(x)\} = \frac{1}{\pi} \int_{-a}^{a} \frac{1}{\sqrt{a^2 - x^2}} e^{-ikx} \, dx$$

$$= \frac{1}{\pi} \int_{-\pi/2}^{\pi/2} e^{-ika \sin\theta} \, d\theta = \frac{1}{2\pi} \int_{-\pi}^{\pi} e^{-ika \sin\theta} \, d\theta$$

and make use of the integral representation of the Bessel function (A.25).

b. General formulae of the Fourier transforms

In the following, the notation of the Fourier transform of $f(x)$ by $F(k) \equiv \mathcal{F}\{f(x)\}$ is used as before.

(i) Multiplication of the variable by a constant

$$\mathcal{F}\{f(ax)\} = \frac{1}{\sqrt{2\pi}} \int_{-\infty}^{\infty} f(ax)e^{-ikx}dx = \frac{1}{a}F\left(\frac{k}{a}\right). \qquad (3.55)$$

(ii) Translation of a given function

$$\mathcal{F}\{f(x-a)\} = \frac{1}{\sqrt{2\pi}} \int_{-\infty}^{\infty} f(x-a)e^{-ikx}dx = e^{-ika}F(k). \qquad (3.56)$$

(iii) Multiplication of an exponential function

$$\mathcal{F}\{e^{iax}f(x)\} = \frac{1}{\sqrt{2\pi}} \int_{-\infty}^{\infty} f(x)e^{-i(k-a)x}dx = F(k-a), \qquad (3.57)$$

from which we have

$$\mathcal{F}\{\cos(ax)f(x)\} = \mathcal{F}\left\{\frac{e^{iax}+e^{-iax}}{2}f(x)\right\} = \frac{1}{2}[F(k-a)+F(k+a)],$$

$$\mathcal{F}\{\sin(ax)f(x)\} = \mathcal{F}\left\{\frac{e^{iax}-e^{-iax}}{2i}f(x)\right\} = \frac{1}{2i}[F(k-a)-F(k+a)].$$

(iv) Multiplication of x^n

Differentiation of both sides of the Fourier transform

$$\mathcal{F}\{f(x)\} = \frac{1}{\sqrt{2\pi}} \int_{-\infty}^{\infty} f(x)e^{-ikx}dx = F(k)$$

with respect to k gives

$$\frac{1}{\sqrt{2\pi}} \int_{-\infty}^{\infty} (-ix)f(x)e^{-ikx}dx = F'(k).$$

By successive differentiations, we have

$$\frac{1}{\sqrt{2\pi}} \int_{-\infty}^{\infty} (-ix)^n f(x)e^{-ikx}dx = F^{(n)}(k),$$

from which we have

$$\mathcal{F}\{xf(x)\} = iF'(k), \qquad \mathcal{F}\{x^n f(x)\} = i^n F^{(n)}(k). \qquad (3.58)$$

(v) Derivatives of a given function

The Fourier transform of the derivative of a given function is

$$\mathcal{F}\left\{\frac{df}{dx}\right\} = \frac{1}{\sqrt{2\pi}} \int_{-\infty}^{\infty} \frac{df}{dx}e^{-ikx}dx$$

$$= \frac{1}{\sqrt{2\pi}} \left([fe^{-ikx}]_{-\infty}^{\infty} + ik \int_{-\infty}^{\infty} f(x)e^{-ikx}dx\right) = ikF(k), \qquad (3.59)$$

and similarly

$$\mathcal{F}\left\{\frac{d^n f}{dx^n}\right\} = (ik)^n F(k), \tag{3.60}$$

where we have assumed f, f', f'', ... $\to 0$ as $x \to \pm\infty$.

(vi) Parseval's relation

A direct calculation of the following gives

$$\int_{-\infty}^{\infty} F(k)\,G(k)\,dk = \int_{-\infty}^{\infty} G(k)\,dk\frac{1}{\sqrt{2\pi}} \int_{-\infty}^{\infty} f(x)\,e^{-ikx}dx$$

$$= \int_{-\infty}^{\infty} f(x)\,dx\frac{1}{\sqrt{2\pi}} \int_{-\infty}^{\infty} G(k)\,e^{-ikx}\,dk$$

$$= \int_{-\infty}^{\infty} f(x)g(-x)\,dx, \tag{3.61}$$

or equivalently

$$\int_{-\infty}^{\infty} F(k)\bar{G}(k)\,dk = \int_{-\infty}^{\infty} f(x)\bar{g}(x)\,dx, \tag{3.62}$$

where the over-bar of a quantity Q, such as $\bar{Q}$, shows the complex conjugate of that quantity Q. In particular, if $f = g$,

$$\int_{-\infty}^{\infty} |F(k)|^2\,dk = \int_{-\infty}^{\infty} |f(x)|^2\,dx. \tag{3.63}$$

The latter relation corresponds to the Parseval's relation in the Fourier series.[3]

(vii) Convolution (or "Faltung" in German)

The **convolution** used in the Fourier transforms is defined by

$$f * g = \int_{-\infty}^{\infty} f(u)g(x-u)\,du, \quad \text{or} \quad \int_{-\infty}^{\infty} f(x-u)g(u)du. \tag{3.64}$$

(The symmetry of f and g will be easily confirmed by the change of variables.) Note the difference in the definition of a convolution in the Laplace transforms, as shown in the later subsection (§§3.3.3,b(vi)).

Denoting the Fourier transforms of the given functions $f(x)$ and $g(x)$ by $\mathcal{F}\{f(x)\} = F(k)$ and $\mathcal{F}\{g(x)\} = G(k)$, respectively, we have

$$\mathcal{F}\{f * g\} = \sqrt{2\pi}F(k)G(k), \quad \mathcal{F}^{-1}\{F(k)G(k)\} = \frac{1}{\sqrt{2\pi}}f * g. \tag{3.65}$$

---

[3]The Parseval's relation in the Fourier series is

$$\frac{1}{\pi} \int_0^{2\pi} \{f(x)\}^2\,dx = \frac{1}{2}a_0^2 + \sum_{n=1}^{\infty}(a_n^2 + b_n^2).$$

This relation is derived by a direct calculation. For example, the l.h.s. of the first relation of Eq. (3.65) is

$$\mathcal{F}\{f * g\} = \frac{1}{\sqrt{2\pi}} \int_{-\infty}^{\infty} \left(\int_{-\infty}^{\infty} f(u)\, g(x-u)\, \mathrm{d}u \right) \mathrm{e}^{-ikx}\, \mathrm{d}x$$

$$= \frac{1}{\sqrt{2\pi}} \int_{-\infty}^{\infty} f(u)\, \mathrm{e}^{-iku} \left(\int_{-\infty}^{\infty} g(x-u)\, \mathrm{e}^{-ik(x-u)}\, \mathrm{d}x \right) \mathrm{d}u$$

$$= \left(\frac{1}{\sqrt{2\pi}} \int_{-\infty}^{\infty} f(u)\mathrm{e}^{-iku}\, \mathrm{d}u \right) \left(\frac{1}{\sqrt{2\pi}} \int_{-\infty}^{\infty} g(v)\, \mathrm{e}^{-ikv}\, \mathrm{d}v \right) \times \sqrt{2\pi}$$

$$= \sqrt{2\pi} F(k)\, G(k).$$

Generalization of the convolution is also useful.[4]

Example 3.4. As a simple example showing the utility of the convolution, we shall solve the equation

$$\left(-\frac{\mathrm{d}^2}{\mathrm{d}x^2} + \alpha^2 \right) \psi = f(x), \tag{3.66}$$

where we choose $\alpha > 0$. By applying the Fourier transform to both sides of the above equation, we have

$$(k^2 + \alpha^2)\Psi(k) = F(k), \tag{3.67}$$

where we denote $\mathcal{F}\{\psi(x)\} = \Psi(k)$ and $\mathcal{F}\{f(x)\} = F(k)$. The solution in the transformed space is easily obtained:

$$\Psi(k) = \frac{F(k)}{k^2 + \alpha^2}, \tag{3.68}$$

which is inversely transformed to obtain the solution in the original space. To this end, we take into account the relation (see §§3.3.2a(iii)-(1))

$$\frac{1}{k^2 + \alpha^2} = \mathcal{F}\left\{ \frac{1}{\alpha}\sqrt{\frac{\pi}{2}}\, \exp(-\alpha|x|) \right\} \equiv G(k), \tag{3.69}$$

and make use of the convolution. Namely,

$$\psi(x) = \frac{1}{\sqrt{2\pi}} \int_{-\infty}^{\infty} F(k)G(k)\mathrm{e}^{ikx}\, \mathrm{d}k = \frac{1}{2\alpha} \int_{-\infty}^{\infty} f(u) \exp(-\alpha|x-u|)\, \mathrm{d}u. \tag{3.70}$$

The last expression also provides a view point that $\exp(-\alpha|x-u|)/2\alpha$ is the Green's function of Eq. (3.66) [see also Table 2.2, column (c), one-dimensional case].   ◁

4

$$\mathcal{F}\left\{ \int_{-\infty}^{\infty} f_n(u_n)\, \mathrm{d}u_n \int_{-\infty}^{\infty} f_{n-1}(u_{n-1})\, \mathrm{d}u_{n-1} ... \int_{-\infty}^{\infty} f_1(u_1)g(x - u_1 - ... - u_n)\, \mathrm{d}u_1 \right\}$$

$$\equiv \mathcal{F}\{f_1 * f_2 * ... * f_n * g\} = (2\pi)^{n/2} F_1(k)F_2(k)...F_n(k)G(k).$$

c. Further examples

Example 3.5. Solve the two-dimensional Laplace's equation

$$\frac{\partial^2 T}{\partial x^2} + \frac{\partial^2 T}{\partial y^2} = 0, \tag{3.71}$$

under the boundary condition

$$T(x,0) = f(x), \quad T(x,b) = 0, \quad (0 \le x < \infty); \quad T(0,y) = 0, \quad (0 \le y \le b). \tag{3.72}$$

This is a problem of obtaining a steady temperature distribution in a semi-infinite strip, where the temperature is given on one side of the boundary, whereas the temperature on the remaining boundary is kept at zero, and hence $f(0) = 0$ is assumed (see Fig. 3.5(a)).

Application of the Fourier sine transform

$$\mathcal{F}_s\{T(x,y)\} \equiv \sqrt{\frac{2}{\pi}} \int_0^\infty T(x,y) \sin(\xi x) \, \mathrm{d}x = \Phi(\xi, y)$$

to Eq. (3.71) yields

$$\frac{\mathrm{d}^2 \Phi}{\mathrm{d}y^2} = \xi^2 \Phi, \tag{3.73}$$

so that we obtain the solution in the transformed space:

$$\Phi = A(\xi) \cosh(\xi y) + B(\xi) \sinh(\xi y),$$

where we have assumed $\partial T/\partial x$, $T \to 0$ as $x \to \infty$. The boundary conditions at $y = 0$, b are

$$A(\xi) = \sqrt{\frac{2}{\pi}} \int_0^\infty f(x) \sin(\xi x) \, \mathrm{d}x = \mathcal{F}_s\{f(x)\},$$

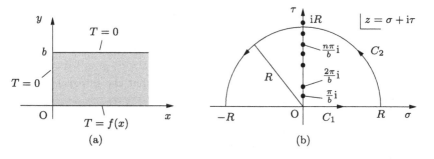

Fig. 3.5 Steady temperature distribution in a semi-infinite strip region.

$$A(\xi)\cosh(\xi b) + B(\xi)\sinh(\xi b) = 0,$$

from which A and B, and hence the solution Φ is determined:

$$\Phi(\xi, y) = \frac{\sinh[\xi(b-y)]}{\sinh(\xi b)}A(\xi). \tag{3.74}$$

The solution $T(x, y)$ is obtained by applying the inverse Fourier sine transform to Φ, which is made as follows:

$$
\begin{aligned}
T(x, y) &= \sqrt{\frac{2}{\pi}} \int_0^\infty \frac{\sinh[\xi(b-y)]}{\sinh(\xi b)} A(\xi) \sin(\xi x)\, d\xi \\
&= \frac{2}{\pi} \int_0^\infty d\xi\, \frac{\sinh[\xi(b-y)]}{\sinh(\xi b)} \sin(\xi x) \int_0^\infty f(x')\sin(\xi x')\, dx' \\
&= \int_0^\infty f(x')\, dx'\, \frac{2}{\pi} \int_0^\infty d\xi\, \frac{\sinh[\xi(b-y)]}{\sinh(\xi b)} \sin(\xi x)\sin(\xi x') \\
&= \int_0^\infty f(x')\, I(x, y, x')\, dx', \tag{3.75}
\end{aligned}
$$

where

$$I(x, y, x') = \frac{2}{\pi} \int_0^\infty \frac{\sinh[\xi(b-y)]}{\sinh(\xi b)} \sin(\xi x)\,\sin(\xi x')\, d\xi. \tag{3.76}$$

Our remaining task is to calculate the integral I. To do this, we extend the variable ξ into a complex variable z, and apply the complex function theory. Taking into account that the integrand is an even function of ξ, we extend the interval of the integral to $(-\infty, \infty)$. Paying our attention to the relation

$$\sin(\xi x)\sin(\xi x') = \frac{1}{2}\{\cos[\xi(x-x')] - \cos[\xi(x+x')]\} = \frac{1}{2}\mathrm{Re}[e^{i\xi(x-x')} - e^{i\xi(x+x')}],$$

we consider the contour integral

$$J = \int_C \frac{\sinh(cz)}{\sinh(bz)} e^{iaz}\, dz,$$

along the contour $C = C_1 + C_2$ as shown in Fig. 3.5(b) (note that $z = 0$ is not a singular point), where $z = \sigma + i\tau$, and we denote $c = b - y$, $a = x \mp x'$ (≥ 0). On the contour C_1, $z = \sigma$, ($\sigma = -R \to R$), so that the integral over this part J_1 is

$$J_1 = \int_{-R}^R \frac{\sinh(c\sigma)}{\sinh(b\sigma)} e^{ia\sigma}\, d\sigma.$$

The real part of the latter with $R \to \infty$ gives $2\pi I$. On the contour C_2, where $z = R\,e^{i\theta}$, ($\theta = 0 \to \pi$), the integral tends to zero as $R \to \infty$ owing to the Jordan's lemma (see *e.g.*, [Brown and Churchill (2009); Fujiwara (2013)]). Meanwhile, singular points inside the closed contour C are $z = (n\pi/b)\,i$, $(n = 1, 2, 3, ...)$, which are all simple poles (order 1). By the use of the residue theorem, the contribution from the pole $z \equiv nz_1 i = (n\pi/b)\,i$ (for brevity we denote $z_1 = \pi/b$) is

$$2\pi i \operatorname{Res}(nz_1 i) = 2\pi i \frac{\sinh(nz_1 ci)}{b \cosh(nz_1 bi)} e^{-z_1 na}$$

$$= \frac{(-1)^n \pi i}{b}[e^{-nz_1(a-ci)} - e^{-nz_1(a+ci)}],$$

so that summing up all contributions from the poles inside C with $R \to \infty$ yields

$$\sum_{n=1}^{\infty} 2\pi i \operatorname{Res}(nz_1 i) = -\frac{\pi i}{b}\left(\frac{e^{-z_1(a-ci)}}{1 + e^{-z_1(a-ci)}} - \frac{e^{-z_1(a+ci)}}{1 + e^{-z_1(a+ci)}}\right)$$

$$= \frac{\pi}{b} \frac{\sin\left(\dfrac{\pi c}{b}\right)}{\cosh\left(\dfrac{\pi a}{b}\right) + \cos\left(\dfrac{\pi c}{b}\right)}.$$

A similar calculation is made for $a = x - x'$ and $a = x + x'$, which yields

$$I(x, y, x') = \frac{\sin\left(\dfrac{\pi y}{b}\right)}{2b}\left(\frac{1}{\cosh\left(\dfrac{\pi(x - x')}{b}\right) + \cos\left(\dfrac{\pi(b - y)}{b}\right)}\right.$$

$$\left. - \frac{1}{\cosh\left(\dfrac{\pi(x + x')}{b}\right) + \cos\left(\dfrac{\pi(b - y)}{b}\right)}\right),$$

$$(3.77)$$

(*cf.* [Titchmarsh (1948)](§10.11)). The present problem is a special case of $a \to \infty$ in §§2.5.1b, where the Dirichlet's problem for the Laplace equation in the rectangular region $(0 \leq x \leq a, 0 \leq y \leq b)$ is dealt with. Indeed, if we rewrite the expression (2.258)

$$\sum_{n=1}^{\infty} \frac{2}{a} \sin\left(\frac{n\pi x}{a}\right) \dots$$

in terms of

$$\frac{\pi}{a} = \Delta\eta, \quad \frac{n\pi}{a} = n\Delta\eta = \eta, \dots$$

and take a limiting process $a \to \infty$:

$$\frac{2}{\pi} \sum_{n=1}^{\infty} \Delta\eta \sin(\eta x) \to \frac{2}{\pi} \int_{0}^{\infty} \sin(\eta x) \ \dots \ \mathrm{d}\eta,$$

we can reproduce Eq. (3.76). ◁

Example 3.6. Obtain the solution of the one-dimensional wave equation

$$\frac{\partial^2 u}{\partial t^2} = c^2 \frac{\partial^2 u}{\partial x^2} \tag{3.78}$$

under the initial condition ($t = 0$)

$$u(x, 0) = f(x), \qquad \frac{\partial}{\partial t} u(x, 0) = g(x). \tag{3.79}$$

The same problem is dealt with in §§2.5.2a, but we shall reconsider here to show the utility of the Fourier transform. The following method that reduces the original PDE to an ODE by applying the Fourier transform to the spatial variable of the former is called the **Stokes' method**.

We first apply the Fourier transform

$$U(k, t) = \frac{1}{\sqrt{2\pi}} \int_{-\infty}^{\infty} u(x, t) \mathrm{e}^{-ikx} \mathrm{d}x, \tag{3.80}$$

to both sides of Eq. (3.78), by which Eq. (3.78) becomes

$$\frac{\mathrm{d}^2 U}{\mathrm{d}t^2} = -c^2 k^2 U. \tag{3.81}$$

This is an ODE of U, and the general solution is

$$U(k, t) = A(k) \cos(kct) + B(k) \sin(kct). \tag{3.82}$$

Here, A and B are arbitrary constants (that may depend on k, but are independent of t) to be determined by the initial condition.

By applying the Fourier transforms to the initial conditions, we have

$$\begin{aligned} U(k, 0) &= \frac{1}{\sqrt{2\pi}} \int_{-\infty}^{\infty} u(x, 0) \, \mathrm{e}^{-ikx} \mathrm{d}x \\ &= \frac{1}{\sqrt{2\pi}} \int_{-\infty}^{\infty} f(x) \, \mathrm{e}^{-ikx} \mathrm{d}x \equiv F(k), \end{aligned} \tag{3.83}$$

$$\begin{aligned} U'(k, 0) &= \frac{1}{\sqrt{2\pi}} \int_{-\infty}^{\infty} \frac{\partial}{\partial t} u(x, 0) \, \mathrm{e}^{-ikx} \mathrm{d}x \\ &= \frac{1}{\sqrt{2\pi}} \int_{-\infty}^{\infty} g(x) \, \mathrm{e}^{-ikx} \mathrm{d}x \equiv G(k), \end{aligned} \tag{3.84}$$

where F and G are the Fourier transforms of f and g, respectively. The general solution (3.82) that satisfies the above-mentioned initial condition is given by $A = F$, $B = G/(kc)$, so that the solution in the transformed space is

$$U(k,t) = F(k)\cos(kct) + \frac{G(k)}{kc}\sin(kct). \tag{3.85}$$

Then, we perform the inverse Fourier transform to Eq. (3.85) to obtain the solution in the original space:

$$
\begin{aligned}
u(x,t) &= \frac{1}{\sqrt{2\pi}}\int_{-\infty}^{\infty} U(k,t)e^{ikx}dk \\
&= \frac{1}{\sqrt{2\pi}}\int_{-\infty}^{\infty}\left[F(k)\cos(kct) + \frac{1}{kc}G(k)\sin(kct)\right]e^{ikx}\,dk.
\end{aligned}
\tag{3.86}
$$

Since, the first term of the r.h.s.

$$
\begin{aligned}
&= \frac{1}{\sqrt{2\pi}}\int_{-\infty}^{\infty} F(k)\cos(kct)\,e^{ikx}dk \\
&= \frac{1}{2\pi}\int_{-\infty}^{\infty}\left(\int_{-\infty}^{\infty} f(x')e^{-ikx'}\,dx'\right)\cos(kct)\,e^{ikx}dk \\
&= \frac{1}{2\pi}\int_{-\infty}^{\infty} f(x')\left(\int_{-\infty}^{\infty} e^{-ikx'}\cos(kct)e^{ikx}dk\right)dx' \\
&= \frac{1}{2}\int_{-\infty}^{\infty} f(x')\left(\frac{1}{2\pi}\int_{-\infty}^{\infty}[e^{ik(x+ct-x')} + e^{ik(x-ct-x')}]\,dk\right)dx' \\
&= \frac{1}{2}\int_{-\infty}^{\infty} f(x')\,[\delta(x+ct-x') + \delta(x-ct-x')]\,dx' \\
&= \frac{1}{2}[f(x+ct) + f(x-ct)],
\end{aligned}
$$

and the second term of the r.h.s.

$$
\begin{aligned}
&= \frac{1}{\sqrt{2\pi}}\int_{-\infty}^{\infty}\frac{1}{kc}G(k)\sin(kct)\,e^{ikx}dk \\
&= \frac{1}{2\pi}\int_{-\infty}^{\infty}\frac{1}{kc}\left(\int_{-\infty}^{\infty} g(x')e^{-ikx'}\,dx'\right)\sin(kct)\,e^{ikx}dk \\
&= \frac{1}{2c}\int_{-\infty}^{\infty} g(x')\left(\frac{1}{2\pi}\int_{-\infty}^{\infty}\frac{e^{ik(x+ct-x')} - e^{ik(x-ct-x')}}{ik}dk\right)dx' \\
&= \frac{1}{2c}\left(\int_{-\infty}^{x+ct} g(x')\,dx' - \int_{-\infty}^{x-ct} g(x')\,dx'\right) = \frac{1}{2c}\int_{x-ct}^{x+ct} g(x')\,dx',
\end{aligned}
$$

we finally have the general solution

$$u(x,t) = \frac{1}{2}[f(x-ct) + f(x+ct)] + \frac{1}{2c}\int_{x-ct}^{x+ct} g(x')dx'. \tag{3.87}$$

In the above calculations, the Fourier transform of the delta function Eq. (2.176) as well as the unit step function Eq. (2.177) has been used.[5] Note that Eq. (3.87) is the same as the d'Alembert–Stokes' solution (2.261).

◁

Example 3.7. Obtain the solution of the one-dimensional diffusion equation

$$\frac{\partial u}{\partial t} = D\frac{\partial^2 u}{\partial x^2} \tag{3.88}$$

under the initial conditions $u(x,0) = f(x)$.

Similar to the previous example, we apply the Fourier transform

$$U(k,t) = \frac{1}{\sqrt{2\pi}} \int_{-\infty}^{\infty} u(x,t)\,e^{-ikx}\,dx \tag{3.89}$$

to both sides of Eq. (3.88) as well as to the initial conditions:

$$\frac{dU(k,t)}{dt} = -Dk^2 U(k,t), \tag{3.90}$$

$$U(k,0) = \frac{1}{\sqrt{2\pi}} \int_{-\infty}^{\infty} f(x)\,e^{-ikx}dx \equiv F(k). \tag{3.91}$$

The solution in the transformed space is

$$U(k,t) = F(k)\,e^{-Dk^2t}, \tag{3.92}$$

which is inversely transformed to obtain our final result:

$$\begin{aligned}
u(x,t) &= \frac{1}{\sqrt{2\pi}} \int_{-\infty}^{\infty} F(k)\,e^{-Dk^2t+ikx}dk \\
&= \frac{1}{2\pi} \int_{-\infty}^{\infty} \left(\int_{-\infty}^{\infty} f(x')\,e^{-ikx'}dx' \right) e^{-Dk^2t+ikx}dk \\
&= \frac{1}{2\pi} \int_{-\infty}^{\infty} f(x') \left(\int_{-\infty}^{\infty} e^{-Dk^2t+ik(x-x')}dk \right) dx' \\
&= \frac{1}{\sqrt{4\pi Dt}} \int_{-\infty}^{\infty} f(x')\exp\left(-\frac{(x-x')^2}{4Dt} \right) dx'.
\end{aligned} \tag{3.93}$$

Here, the integral

$$\int_{-\infty}^{\infty} e^{-ax^2+ibx}dx = \sqrt{\frac{\pi}{a}}\exp\left(-\frac{b^2}{4a} \right), \quad (a > 0,\ b : \text{real})$$

is assumed to be known [Brown and Churchill (2009); Fujiwara (2013)]. Refer also to the Green's function Eqs. (2.244) [or (2.274)]. See also Table 2.2, column (e), one-dimensional case.

◁

5

$$\frac{1}{2\pi} \int_{-\infty}^{\infty} e^{ikx}dk = \delta(x), \qquad \frac{1}{2\pi} \int_{-\infty}^{\infty} \frac{e^{ikx}}{ik}dk = \begin{cases} 0 & (x < 0) \\ 1 & (x > 0) \end{cases}.$$

3.3.3 Laplace transforms

The Laplace transforms have been fully explained in many books *e.g.*, [Morse and Feshbach (1953); Erdélyi (1954); Abramowitz and Stegun (1964)]. Here, we restrict our attention to the relation to the Fourier transforms. Only a few simple but important examples are given to show their utility.

We first extend the variable s into a complex variable $s = x + iy$ in the Laplace transform:

$$U(s) = \int_0^\infty e^{-st} u(t)\, dt.$$

Then, it follows that

$$U(x + iy) = \int_0^\infty e^{-(x+iy)t} u(t)\, dt = \int_0^\infty e^{-xt} u(t)\, e^{-iyt}\, dt.$$

By defining

$$v(t) = \begin{cases} \sqrt{2\pi}\, e^{-xt} u(t) & (t \geq 0) \\ 0 & (t < 0) \end{cases},$$

we have

$$U(x + iy) = \frac{1}{\sqrt{2\pi}} \int_{-\infty}^\infty v(t)\, e^{-iyt}\, dt,$$

which is the Fourier transform (3.37) itself. Application of the inverse Fourier transform gives

$$u(t) = \frac{e^{xt}}{\sqrt{2\pi}}\, v(t) = \frac{e^{xt}}{2\pi} \int_{-\infty}^\infty U(x + iy)\, e^{iyt}\, dy.$$

If we put $x + iy = z$, then the contour in the complex z plane corresponding to $-\infty < y < \infty$ becomes $\gamma - i\infty < z < \gamma + i\infty$ $(0 < \gamma < 1)$, so that we have

$$u(t) = \frac{1}{2\pi i} \int_{\gamma - i\infty}^{\gamma + i\infty} U(z)\, e^{zt}\, dz,$$

which yields the inverse Laplace transform (3.46).

a. Simple examples

We show some fundamental properties of the Laplace transforms.
(i) We start with the integral

$$\mathcal{L}\{e^{ax}\} = \int_0^\infty e^{-(s-a)x}\, dx = \frac{1}{s-a}, \tag{3.94}$$

where we assume the convergence of the integral $(s > a)$.

(ii) We differentiate the above relation with respect to a (here, parameter a is assumed to be continuous):

$$\mathcal{L}\{xe^{ax}\} = \frac{1}{(s-a)^2}, \ \mathcal{L}\{x^2 e^{ax}\} = \frac{2}{(s-a)^3}, \ \cdots, \ \mathcal{L}\{x^n e^{ax}\} = \frac{n!}{(s-a)^{n+1}}. \tag{3.95}$$

(iii) By putting $a = 0$ in the above relation, we have

$$\mathcal{L}\{1\} = \frac{1}{s}, \quad \mathcal{L}\{x\} = \frac{1}{s^2}, \quad \mathcal{L}\{x^2\} = \frac{2}{s^3}, \quad \cdots, \quad \mathcal{L}\{x^n\} = \frac{n!}{s^{n+1}}. \tag{3.96}$$

(iv) The above relations hold for complex variable a (convergence of the integral is assumed), so that

$$\mathcal{L}\{e^{i\omega x}\} = \frac{1}{s - i\omega}, \tag{3.97}$$

from which we have

$$\mathcal{L}\{\cos(\omega x)\} = \mathcal{L}\left\{\frac{e^{i\omega x} + e^{-i\omega x}}{2}\right\} = \frac{1}{2}\left(\frac{1}{s - i\omega} + \frac{1}{s + i\omega}\right) = \frac{s}{s^2 + \omega^2}, \tag{3.98}$$

$$\mathcal{L}\{\sin(\omega x)\} = \mathcal{L}\left\{\frac{e^{i\omega x} - e^{-i\omega x}}{2i}\right\} = \frac{1}{2i}\left(\frac{1}{s - i\omega} - \frac{1}{s + i\omega}\right) = \frac{\omega}{s^2 + \omega^2}. \tag{3.99}$$

Further differentiation of the above relation with respect to ω yields

$$\mathcal{L}\{x\sin(\omega x)\} = \frac{2\omega s}{(s^2 + \omega^2)^2}, \quad \mathcal{L}\{x\cos(\omega x)\} = \frac{s^2 - \omega^2}{(s^2 + \omega^2)^2}, \tag{3.100}$$

and

$$\mathcal{L}\{e^{\gamma x}\cos(\omega x)\} = \mathcal{L}\left\{\frac{e^{(\gamma + i\omega)x} + e^{(\gamma - i\omega)x}}{2}\right\} = \cdots = \frac{s - \gamma}{(s - \gamma)^2 + \omega^2}. \tag{3.101}$$

(v) Furthermore, a combination of the above results with the power series expansion of the Bessel function (A.13) yields

$$\mathcal{L}\{J_0(kx)\} = \sum_{n=0}^{\infty} \frac{(-1)^n}{(n!)^2} \int_0^\infty \left(\frac{kx}{2}\right)^{2n} e^{-sx}\, dx = \sum_{n=0}^{\infty} \frac{(-1)^n (2n)!}{(n!)^2 2^{2n} s}\left(\frac{k}{s}\right)^{2n}$$

$$= \sum_{n=0}^{\infty} \frac{(-1)^n (2n-1)!!}{n!\, 2^n\, s}\left(\frac{k}{s}\right)^{2n} = \frac{1}{\sqrt{s^2 + k^2}}. \tag{3.102}$$

The inverse Laplace transforms are performed by calculating Eq. (3.46). An alternative, and much simpler, method is to prepare the table of the Laplace and its inverse transforms, and find the relevant expressions. See e.g. [Erdélyi (1954)](Chaps. IV, V), [Abramowitz and Stegun (1964)](Chap. 29).

b. General formulae of the Laplace transforms

The following general formulae will provide further examples of the Laplace transforms.

(i) Multiplication of the variable by a constant

$$\mathcal{L}\{f(ax)\} = \int_0^\infty f(ax)\,e^{-sx}\,dx$$

$$= \frac{1}{a}\int_0^\infty f(ax)\exp\left(-\frac{s}{a}ax\right)d(ax) = \frac{1}{a}F\left(\frac{s}{a}\right). \qquad (3.103)$$

(ii) Translation of a given function

$$\mathcal{L}\{f(x-a)\} = \int_0^\infty f(x-a)\,e^{-sx}\,dx$$

$$= \int_0^\infty f(x-a)\,e^{-s(x-a)-sa}\,dx = e^{-sa}F(s), \qquad (3.104)$$

where we have assumed $f(x) = 0$ for $x < 0$.

By the use of this formula, the Laplace transform of a periodic function is easily performed. As depicted in Fig. 3.6, a periodic function of period T defined in the range $t \geq 0$ is rewritten as

$$f(t) = f_0(t) + f(t-T), \qquad (3.105)$$

where the function $f_0(t)$ is defined in the interval $[0, T]$. By applying the Laplace transform to both sides of Eq. (105), we have

$$F(s) = F_0(s) + e^{-sT}F(s), \qquad (3.106)$$

from which we have

$$F(s) = \frac{F_0(s)}{1 - e^{-sT}}, \qquad (3.107)$$

where $F_0(s)$ is the Laplace transform of $f_0(t)$:

$$F_0(s) = \mathcal{L}\{f_0(t)\} = \int_0^T f(t)\,e^{-st}\,dt. \qquad (3.108)$$

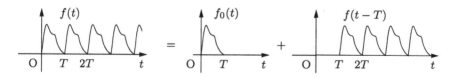

Fig. 3.6   Periodic function.

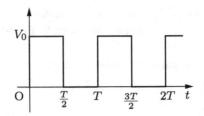

Fig. 3.7   Clock signal (rectangular wave).

Example 3.8. Calculate the Laplace transform of the function as shown in Fig. 3.7.

Here, $f(t)$ is a periodic function of period T, within which it is defined by

$$f_0(t) = \begin{cases} V_0 & (0 < t < T/2) \\ 0 & (T/2 < t < T) \end{cases} . \tag{3.109}$$

The Laplace transform of $f_0(t)$ in the first period is

$$F_0(s) = \int_0^{T/2} V_0\, e^{-st} dt = \frac{V_0}{s}\left(1 - e^{-sT/2}\right), \tag{3.110}$$

so that the Laplace transform of $f(t)$ is

$$F(s) = \frac{V_0(1 - e^{-sT/2})}{s(1 - e^{-sT})} = \frac{V_0}{s(1 + e^{-sT/2})}. \tag{3.111}$$

$\lhd$

(iii) Multiplication of an exponential function

By a direct calculation, we have

$$\mathcal{L}\{e^{ax} f(x)\} = \int_0^\infty f(x)\, e^{-(s-a)x}\, dx = F(s - a). \tag{3.112}$$

Using this formula, we can also obtain the previously obtained relation [Eq. (3.101)]:

$$\mathcal{L}\{e^{\gamma x} \cos(\omega x)\} = \left.\frac{s}{s^2 + \omega^2}\right|_{s \to s - \gamma} = \frac{s - \gamma}{(s - \gamma)^2 + \omega^2}.$$

(iv) Multiplication of x^n

By a direct calculation, we have

$$F'(s) = \frac{d}{ds} \int_0^\infty f(x)\, e^{-sx}\, dx = \int_0^\infty (-x)\, f(x)\, e^{-sx}\, dx = \mathcal{L}\{-x f(x)\}. \tag{3.113}$$

In general, we have

$$F^{(n)}(s) = \frac{\mathrm{d}^n}{\mathrm{d}s^n} \int_0^\infty f(x)\,\mathrm{e}^{-sx}\,\mathrm{d}x = \int_0^\infty (-x)^n f(x)\,\mathrm{e}^{-sx}\mathrm{d}x = \mathcal{L}\{(-x)^n f(x)\}.$$
(3.114)

(v) Integrals of a given function

By a direct calculation, we have

$$\mathcal{L}\left\{\int_0^x f(u)\,\mathrm{d}u\right\} = \int_0^\infty \mathrm{e}^{-sx}\left(\int_0^x f(u)\,\mathrm{d}u\right)\mathrm{d}x$$

$$= \left[-\frac{\mathrm{e}^{-sx}}{s}\int_0^x f(u)\,\mathrm{d}u\right]_0^\infty + \frac{1}{s}\int_0^x \mathrm{e}^{-sx}f(x)\,\mathrm{d}x = \frac{1}{s}F(s).$$
(3.115)

(vi) Convolution

The **convolution** used in the Laplace transforms is defined by

$$f * g = \int_0^x f(x-u)\,g(u)\,\mathrm{d}u = \int_0^x f(u)\,g(x-u)\,\mathrm{d}u.$$
(3.116)

By applying the Laplace transform, we have

$$\mathcal{L}\{f * g\} = \int_0^\infty \mathrm{e}^{-sx}\left(\int_0^x f(x-u)\,g(u)\,\mathrm{d}u\right)\mathrm{d}x$$

$$= \int_0^\infty g(u)\left(\int_x^\infty f(x-u)\,\mathrm{e}^{-sx}\mathrm{d}x\right)\mathrm{d}u$$

$$= \int_0^\infty g(u)\,\mathrm{e}^{-su}\left(\int_0^\infty f(v)\,\mathrm{e}^{-sv}\mathrm{d}v\right)\mathrm{d}u = F(s)\,G(s),$$

and hence[6]

$$\mathcal{L}\{f * g\} = F(s)G(s), \quad \mathcal{L}^{-1}\{F(s)G(s)\} = f * g.$$
(3.117)

(vii) Derivatives of a given function

By a direct calculation, we have

$$\mathcal{L}\{u'(x)\} = \int_0^\infty u'(x)\,\mathrm{e}^{-sx}\mathrm{d}x = \left[u(x)\,\mathrm{e}^{-sx}\right]_0^\infty + s\int_0^\infty u(x)\,\mathrm{e}^{-sx}\mathrm{d}x$$

$$= s\mathcal{L}\{u(x)\} - u(0) = sU(s) - u(0).$$
(3.118)

The repeated use of this formula yields

$$\mathcal{L}\{u''(x)\} = s\mathcal{L}\{u'(x)\} - u'(0) = s^2\mathcal{L}\{u(x)\} - [su(0) + u'(0)], \quad (3.119)$$

$$\mathcal{L}\{u^{(n)}(x)\} = s^n\mathcal{L}\{u(x)\} - [s^{n-1}u(0) + s^{n-2}u'(0) + ... + u^{(n-1)}(0)]. \quad (3.120)$$

---

[6]In the derivation, we have changed the order of integration with respect to u and x in the second to third terms, and then put $x - u = v$ in the last integration.

Taking account of these properties, we apply the Laplace transform to the inhomogeneous n-th order linear differential equation:

$$a_n y^{(n)}(x) + \ldots + a_1 y'(x) + a_0 y(x) = f(x), \quad a_n \neq 0. \qquad (3.121)$$

Then, we have the following algebraic equation:

$$a_n s^n Y(s) - a_n [s^{n-1} y(0) + s^{n-2} y'(0) + \cdots + y^{(n-1)}(0)]$$
$$+ a_{n-1} s^{n-1} Y(s) - a_{n-1} [s^{n-2} y(0) + s^{n-3} y'(0) + \cdots + y^{(n-2)}(0)]$$
$$+ \cdots + a_1 s Y(s) - a_1 y(0) + a_0 Y(s) = F(s),$$

where we denote $\mathcal{L}\{y(x)\} = Y(s)$ and $\mathcal{L}\{f(x)\} = F(s)$. Accordingly, the above equation is expressed in the following form:

$$\phi(s) Y(s) = F(s) + P_{n-1}(s), \qquad (3.122)$$

where

$$\phi(s) = a_n s^n + a_{n-1} s^{n-1} + \ldots + a_1 s + a_0 \qquad (3.123)$$

is called the **characteristic polynomial**, and

$$P_{n-1}(s) = a_n [s^{n-1} y(0) + s^{n-2} y'(0) + \ldots + y^{(n-1)}(0)] + \ldots + a_1 y(0) \quad (3.124)$$

is an $(n-1)$-th order polynomial depending on the initial condition.

In this way, the differential equation (ODE) in the original space is reduced to the algebraic equation in the transformed space, so that the solution of the latter is easily obtained:

$$Y(s) = \frac{F(s)}{\phi(s)} + \frac{P_{n-1}(s)}{\phi(s)}. \qquad (3.125)$$

By applying the inverse Laplace transform $y(x) = \mathcal{L}^{-1}\{Y(s)\}$, the solution in the original space is obtained. As shown above, the Laplace transform is often used to solve the initial value problems.

Note that the method used to solve Eq. (3.125) by applying inverse Laplace transform is essentially the same as the Mikusiński's operational method.

c. Further examples

Example 3.9. Let us consider the two-port network (4-terminal circuit) as shown in Fig. 3.8(a), and obtain the output $e_2(t)$ for the input signal $e_1(t)$, where R is the resistance, and C is the capacitance of the capacitor.

By choosing the direction of current $i(t)$ as shown in Fig. 3.8(a), the Kirchhoff's second law gives

$$e_1 = R i + \frac{1}{C} \int_0^t i \, dt, \quad e_2 = \frac{1}{C} \int_0^t i \, dt. \qquad (3.126)$$

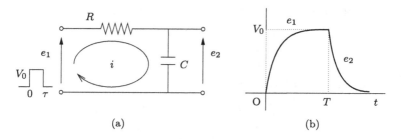

Fig. 3.8   Transient response in two-port network (4-terminal circuit).

Application of the Laplace transforms to both sides of these equations, we have

$$E_1 = RI + \frac{1}{Cs}I, \quad E_2 = \frac{1}{Cs}I, \tag{3.127}$$

where we denote $E_1 = \mathcal{L}\{e_1\}, E_2 = \mathcal{L}\{e_2\}, I = \mathcal{L}\{i\}$. The solution in the transformed space is easily obtained

$$I = \frac{E_1}{R + \dfrac{1}{Cs}}, \quad E_2 = \frac{E_1}{RC\left(s + \dfrac{1}{RC}\right)}. \tag{3.128}$$

The response $e_2(t)$ is obtained either by (1) calculating $E_1(s)$ of the Laplace transform of $e_1(t)$, which is substituted for the r.h.s. of $E_2(s)$, and then applying the inverse Laplace transform to $E_2(s)$, or (2) making use of the convolution formula mentioned in the previous subsection.

In the latter, we note the relation

$$E_2 = \frac{E_1}{RC\left(s + \dfrac{1}{RC}\right)} = \frac{1}{RC}\mathcal{L}\{e_1(t)\}\mathcal{L}\{e^{-t/RC}\}, \tag{3.129}$$

and apply the convolution formula (3.117), from which we have

$$e_2(t) = \frac{1}{RC}\int_0^t e_1(u)\,e^{-(t-u)/RC}du. \tag{3.130}$$

In particular, if the input is given by

$$e_1(t) = \begin{cases} V_0 & (0 \le t \le T) \\ 0 & (t > T) \end{cases}, \tag{3.131}$$

then the output is

$$e_2(t) = \begin{cases} V_0(1 - e^{-t/RC}) & (0 \le t \le T) \\ V_0(e^{-(t-T)/RC} - e^{-t/RC}) & (t > T) \end{cases}. \tag{3.132}$$

We show the behavior of e_2 in Fig. 3.8(b) (solid line). Such a time-dependent behavior is generally called a **transient phenomena**.   ◁

Example 3.10. Obtain the solution of the one-dimensional diffusion equation

$$\frac{\partial u}{\partial t} = \kappa \frac{\partial^2 u}{\partial x^2}, \qquad (\kappa > 0), \tag{3.133}$$

under the initial and boundary conditions:

$$u(x, t) = 0 \qquad (t \le 0), \tag{3.134}$$
$$u(0, t) = T_0 \qquad (t > 0). \tag{3.135}$$

This is a problem of determining the temperature distribution in a thermally insulated semi-infinite cylinder that is initially kept at a temperature of zero, whose end $x = 0$ touches a heat bath of a constant temperature T_0 at $t > 0$.

By applying the Laplace transform

$$U(x, s) = \int_0^\infty u(x, t) \, e^{-st} dt, \tag{3.136}$$

to both sides of Eq. (3.133), we have

$$\kappa \frac{d^2 U}{dx^2} = sU, \tag{3.137}$$

whose solution that satisfies the initial and boundary conditions is

$$U(x, s) = \frac{T_0}{s} \exp\left(-\sqrt{\frac{s}{\kappa}} x \right). \tag{3.138}$$

To obtain the solution in the original space, we shall apply the inverse Laplace transform by making use of the **Bromwich integral**

$$u(x, t) = \frac{1}{2\pi i} \int_{\gamma - i\infty}^{\gamma + i\infty} U(x, s) \, e^{st} ds = \mathcal{L}^{-1}\{U(x, s)\}. \tag{3.139}$$

To perform the calculation, however, caution is required because the integrand given by Eq. (3.138) includes a function $\sqrt{s}$. Since the latter is a multi-valued function with a branch point $s = 0$, we need to introduce a **cut** that connects the branch point and some point at infinity, and restrict the complex plane to a simply connected region. By choosing the latter at minus infinity on the real axis, we consider the contour C as shown in Fig. 3.9. Here, $0 < \gamma < 1$, and C_R and C_R' are the quarter-circles of radius $R(\gg 1)$ in the left-half plane with their centers at the origin (plus short segments at their right ends $\mathrm{Re}\, s \pm iR$ ($0 \le \mathrm{Re}\, s \le \gamma$), which are connected by the line $\mathrm{Re}\, s = \gamma$. We also consider a circle of radius ϵ ($\ll 1$) centered at the origin (C_ϵ), as well as the contours AB and CD along the cut to form a closed contour.

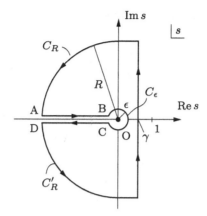

Fig. 3.9   Contour with a cut.

With the above-mentioned contour, the integral along C is zero, because no singularity exists inside the contour (the Cauchy's theorem):

$$\oint_C \cdots \mathrm{d}s = \int_{\gamma-iR}^{\gamma+iR} \cdots \mathrm{d}s + \int_{C_R} \cdots \mathrm{d}s$$
$$+ \int_A^B \cdots \mathrm{d}s + \int_{C_\varepsilon} \cdots \mathrm{d}s + \int_C^D \cdots \mathrm{d}s + \int_{C_{R'}} \cdots \mathrm{d}s = 0,$$

(for details, see *e.g.*, [Brown and Churchill (2009); Fujiwara (2013)]). Then, we have

$$\int_{\gamma-iR}^{\gamma+iR} \cdots \mathrm{d}s = - \left(\int_{C_R} \cdots \mathrm{d}s + \int_A^B \cdots \mathrm{d}s + \int_{C_\varepsilon} \cdots \mathrm{d}s + \int_C^D \cdots \mathrm{d}s + \int_{C_R'} \cdots \mathrm{d}s \right).$$

The integrals along the contours C_R and C_R' become zero as $R \to \infty$, owing to the same argument as developed in deriving the Jordan's lemma. Taking into account that $s = re^{\pi i}$ on the contour AB, so that $\sqrt{s} = i\sqrt{r}$, $(r = R \to \varepsilon)$, whereas $s = re^{-\pi i}$ on the contour CD, so that $\sqrt{s} = -i\sqrt{r}$, $(r = \varepsilon \to R)$, the second and fourth terms of the r.h.s. become

$$-2i \int_\varepsilon^R \frac{1}{r} e^{-rt} \sin\left(\sqrt{\frac{r}{\kappa}} x \right) \mathrm{d}r.$$

The third term of the r.h.s. is $2\pi i$ by the use of the residue theorem.

All included, we have

$$u(x,t) = \frac{1}{2\pi i} \int_{\gamma-iR}^{\gamma+iR} \cdots \mathrm{d}s = T_0 \left[1 - \frac{1}{\pi} \int_0^\infty \frac{1}{r} e^{-rt} \sin\left(\sqrt{\frac{r}{\kappa}} x \right) \mathrm{d}r \right],$$

$$(3.140)$$

as $R \to \infty, \epsilon \to 0$. The above expression is rewritten by a change of variable $\sqrt{r/\kappa} = \xi$:

$$\frac{u}{T_0} = 1 - \frac{2}{\pi} \int_0^\infty e^{-\kappa t \xi^2} \frac{\sin(x\xi)}{\xi} \, d\xi. \tag{3.141}$$

Furthermore, if we remark the integral (see §§3.3.2a(i)-(2))

$$\int_0^\infty e^{-\lambda \xi^2} \cos(x\xi) \, d\xi = \frac{1}{2} \sqrt{\frac{\pi}{\lambda}} \exp\left(-\frac{x^2}{4\lambda}\right),$$

and its integration with respect to x

$$\int_0^\infty e^{-\lambda \xi^2} \frac{\sin(x\xi)}{\xi} \, d\xi = \frac{1}{2} \sqrt{\frac{\pi}{\lambda}} \int_0^x \exp\left(-\frac{x^2}{4\lambda}\right) dx,$$

we have

$$\frac{u}{T_0} = 1 - \frac{1}{\sqrt{\pi \kappa t}} \int_0^x \exp\left(-\frac{x^2}{4\kappa t}\right) dx = 1 - \frac{2}{\sqrt{\pi}} \int_0^{x/\sqrt{4\kappa t}} e^{-\eta^2} \, d\eta. \tag{3.142}$$

In terms of the **error function** $\mathrm{erf}(x)$, and the **complementary error function** $\mathrm{erfc}(x)$ (see Eq. (2.281)), the last expression is further rewritten as

$$u(x,t) = T_0 \left[1 - \mathrm{erf}\left(\frac{x}{2\sqrt{\kappa t}}\right)\right] = T_0 \, \mathrm{erfc}\left(\frac{x}{2\sqrt{\kappa t}}\right). \tag{3.143}$$

◁

3.3.4 Mellin transforms

The Mellin transform is formally derived by extending the interval of the integral in the Laplace transform to $(-\infty, \infty)$. Namely, we consider a function $u(t)$ such that $|u(t)|$ converges sufficiently rapidly to zero as $t \to \pm\infty$, so that the integral

$$U(s) = \int_{-\infty}^\infty e^{-st} u(t) \, dt \tag{3.144}$$

exists. By changing the variable $e^{-t} = x$, the integral (3.144) becomes

$$U(s) = \int_0^\infty u(x) \, x^{s-1} \, dx \equiv \mathcal{M}\{u(x)\}. \tag{3.145}$$

The inverse transform is similarly calculated by

$$u(t) = \frac{1}{2\pi i} \int_{\gamma-i\infty}^{\gamma+i\infty} U(s) e^{st} \, ds = \frac{1}{2\pi i} \int_{\gamma-i\infty}^{\gamma+i\infty} x^{-s} U(s) \, ds \equiv u(x), \tag{3.146}$$

for $0 < \gamma < 1$, which agrees with the definition of the Mellin transform (3.48) and its inverse transform (3.49) mentioned before.

a. Simple examples

By a direct calculation of the integral (3.145), we have

(i) $\quad f(x) = \begin{cases} 1 & (0 \le x \le a) \\ 0 & (x > a) \end{cases} \quad \rightarrow \quad \mathcal{M}\{f(x)\} = \int_0^a x^{s-1}\, dx = \dfrac{a^s}{s}.$

(ii)

$$f(x) = \begin{cases} \log(a/x) & (0 < x \le a) \\ 0 & (x > a) \end{cases} \rightarrow \mathcal{M}\{f(x)\} = \int_0^a x^{s-1} \log \frac{a}{x}\, dx = \frac{a^s}{s^2}.$$

(iii)

$$f(x) = \begin{cases} x^\nu & (0 < x \le a) \\ 0 & (x > a) \end{cases} \rightarrow \mathcal{M}\{f(x)\} = \frac{a^{s+\nu}}{s+\nu}, \quad (\operatorname{Re}(s+\nu) > 0).$$

(iv)

$$\mathcal{M}\left\{\frac{1}{a+x}\right\} = \int_0^\infty \frac{x^{s-1}}{a+x}\, dx = \frac{\pi a^{s-1}}{\sin(\pi s)}, \quad (|\arg a| < \pi;\ 0 < \operatorname{Re} s < 1).$$

(v)

$$\mathcal{M}\left\{\frac{1}{x^2+a^2}\right\} = \int_0^\infty \frac{x^{s-1}}{x^2+a^2}\, dx = \frac{\pi a^{s-2}}{2\sin(\pi s/2)}, \quad (\operatorname{Re} a > 0;\ 0 < \operatorname{Re} s < 2).$$

(vi) $\quad \mathcal{M}\{e^{-ax}\} = a^{-s}\Gamma(s), \quad (\operatorname{Re} a > 0;\ \operatorname{Re} s > 0),$
where $\Gamma(s)$ is the gamma function.

(vii)

$$\mathcal{M}\{\sin(ax)\} = a^{-s}\Gamma(s)\sin\left(\frac{\pi s}{2}\right), \quad (a > 0;\ -1 < \operatorname{Re} s < 1).$$

$$\mathcal{M}\{\cos(ax)\} = a^{-s}\Gamma(s)\cos\left(\frac{\pi s}{2}\right), \quad (a > 0;\ 0 < \operatorname{Re} s < 1).$$

The calculation of the latter has been shown in the example of the Fourier cosine transforms (§§3.3.2a(i)-(3)).

b. General formulae of the Mellin transforms

Mellin transform (3.48)

$$\mathcal{M}\{\phi(x)\} = \int_0^\infty \phi(x)\, x^{s-1}\, dx \equiv \Phi(s),$$

have the following general features.

(i) Multiplication of the variable by a constant

$$\mathcal{M}\{\phi(ax)\} = \int_0^\infty \phi(ax)\, x^{s-1}\, \mathrm{d}x = a^{-s}\Phi(s), \quad a > 0. \qquad (3.147)$$

(ii) Functions of a variable multiplied by its power

$$\mathcal{M}\{x^b\phi(x)\} = \int_0^\infty x^b\, \phi(x)\, x^{s-1}\mathrm{d}x = \Phi(s+b). \qquad (3.148)$$

(iii) Functions given by the power of a variable

$$\mathcal{M}\{\phi(x^c)\} = \int_0^\infty \phi(x^c)\, x^{s-1}\, \mathrm{d}x = \frac{1}{c}\Phi\left(\frac{s}{c}\right), \quad c > 0.$$

$$\mathcal{M}\{\phi(x^{-c})\} = \frac{1}{c}\Phi\left(-\frac{s}{c}\right), \quad c > 0.$$

In particular, if $c = 1$ in the latter, we have

$$\mathcal{M}\left\{\phi\left(\frac{1}{x}\right)\right\} = \Phi(-s). \qquad (3.149)$$

Example 3.11. A combination of the present subsection **b** (i)$\sim$(iii) yields

$$\mathcal{M}\{\phi(ax^c)\} = \frac{1}{c}a^{-s/c}\Phi\left(\frac{s}{c}\right), \quad \mathcal{M}\{x^b\phi(ax^c)\} = \frac{1}{c}a^{-(s+b)/c}\Phi\left(\frac{s+b}{c}\right)$$

for $c > 0$. A similar expression is given for $c < 0$.   ◁

(iv) Derivatives of a given function

$$\mathcal{M}\left\{\frac{\mathrm{d}\phi}{\mathrm{d}x}\right\} = \int_0^\infty \frac{\mathrm{d}\phi}{\mathrm{d}x}\, x^{s-1}\, \mathrm{d}x$$

$$= \left[\phi\, x^{s-1}\right]_0^\infty - (s-1)\int_0^\infty \phi\, x^{s-2}\, \mathrm{d}x = -(s-1)\,\Phi(s-1),$$

and similarly

$$\mathcal{M}\left\{\frac{\mathrm{d}^2\phi}{\mathrm{d}x^2}\right\} = (s-1)(s-2)\,\Phi(s-2), \quad \cdots\, . \qquad (3.150)$$

(v) Euler-type derivatives

$$\mathcal{M}\left\{x\frac{\mathrm{d}\phi}{\mathrm{d}x}\right\} = \int_0^\infty \frac{\mathrm{d}\phi}{\mathrm{d}x}\, x^s\, \mathrm{d}x = \left[\phi\, x^s\right]_0^\infty - s\int_0^\infty \phi\, x^{s-1}\, \mathrm{d}x = -s\,\Phi(s),$$

$$\mathcal{M}\left\{x\frac{\mathrm{d}}{\mathrm{d}x}\left(x\frac{\mathrm{d}\phi}{\mathrm{d}x}\right)\right\} = s^2\,\Phi(s),$$

and in general

$$\mathcal{M}\left\{\left(x\frac{\mathrm{d}}{\mathrm{d}x}\right)^n \phi(x)\right\} = (-1)^n s^n \, \Phi(s), \qquad (3.151)$$

where ϕ, $\mathrm{d}\phi/\mathrm{d}x$, $\mathrm{d}^2\phi/\mathrm{d}x^2,\ldots \to 0$ as $x \to \infty$ is assumed in (iv) and (v).

(vi) Parseval's relation

By a direct calculation, we have

$$\frac{1}{2\pi i}\int_{\gamma-i\infty}^{\gamma+i\infty} \Phi(s)\,\Psi(1-s)\,\mathrm{d}s = \frac{1}{2\pi i}\int_{\gamma-i\infty}^{\gamma+i\infty} \Psi(1-s)\,\mathrm{d}s \int_0^\infty \phi(x)x^{s-1}\,\mathrm{d}x$$

$$= \int_0^\infty \phi(x)\,\mathrm{d}x \frac{1}{2\pi i}\int_{\gamma-i\infty}^{\gamma+i\infty} \Psi(1-s)x^{-(1-s)}\,\mathrm{d}s = \int_0^\infty \phi(x)\,\psi(x)\,\mathrm{d}x.$$

$$(3.152)$$

By choosing $\Psi = \Phi$ (real) and $\gamma = 1/2$, so that $s = 1/2 + it$, we have

$$\frac{1}{2\pi}\int_{-\infty}^\infty \left|\Phi\left(\frac{1}{2}+it\right)\right|^2 \mathrm{d}t = \int_0^\infty \phi(\xi)^2\,\mathrm{d}\xi, \qquad (3.153)$$

which is called the **Parseval's formula** in the Mellin transforms. Similarly we have, by using Eq. (3.149),

$$\frac{1}{2\pi i}\int_{\gamma-i\infty}^{\gamma+i\infty} \Phi(s)\,\Psi(s)\,\mathrm{d}s = \int_0^\infty \frac{1}{\xi}\phi\left(\frac{1}{\xi}\right)\psi(\xi)\,\mathrm{d}\xi. \qquad (3.154)$$

A direct calculation also yields the following relation:

$$\mathcal{M}\left\{x^\alpha \int_0^\infty \xi^\beta\,\phi(x\xi)\,\psi(\xi)\,\mathrm{d}\xi\right\} = \Phi(s+\alpha)\,\Psi(1-s-\alpha+\beta), \qquad (3.155)$$

$$\mathcal{M}\left\{x^\alpha \int_0^\infty \xi^\beta\,\phi\left(\frac{x}{\xi}\right)\psi(\xi)\,\mathrm{d}\xi\right\} = \Phi(s+\alpha)\,\Psi(1+s+\alpha+\beta). \qquad (3.156)$$

For the derivation of Eqs. (3.155) and (3.156), we may exchange the order of integrals, and put $x\xi$ and x/ξ, respectively, as a new variable.[7]

---

[7]Let the l.h.s. of Eq. (3.155) by I, then

$$I = \int_0^\infty \mathrm{d}x\, x^{s-1}\left(x^\alpha \int_0^\infty \xi^\beta\,\phi(x\xi)\,\psi(\xi)\,\mathrm{d}\xi\right) = \int_0^\infty \mathrm{d}\xi\,\xi^\beta\psi(\xi)\int_0^\infty \mathrm{d}x\, x^{s+\alpha-1}\phi(x\xi).$$

By putting $x\xi = \eta$, the integral with respect to x in the second integral becomes

$$\xi^{-(s+\alpha)}\int_0^\infty \eta^{s+\alpha-1}\phi(\eta)\,\mathrm{d}\eta,$$

so that we have

$$I = \int_0^\infty \eta^{s+\alpha-1}\phi(\eta)\,\mathrm{d}\eta \int_0^\infty \xi^{\beta-(s+\alpha)}\psi(\xi)\,\mathrm{d}\xi = \Phi(s+\alpha)\,\Psi(1-s-\alpha+\beta).$$

Relation (3.156) is derived in a similar calculation with the change of variable $x/\xi = \eta$.

By the use of these formulae, we can obtain further general formulae. In particular, we have

$\alpha = \beta = 0$ in Eq. (3.155):

$$\mathcal{M}\left\{\int_0^\infty \phi(x\xi)\,\psi(\xi)\,\mathrm{d}\xi\right\} = \Phi(s)\Psi(1-s), \qquad (3.157)$$

$\alpha = \beta = 0$ in Eq. (3.156):

$$\mathcal{M}\left\{\int_0^\infty \phi\left(\frac{x}{\xi}\right)\psi(\xi)\,\mathrm{d}\xi\right\} = \Phi(s)\Psi(1+s). \qquad (3.158)$$

(vii) Convolution

By putting $\alpha = 0$, $\beta = -1$ in Eq. (3.156), we have

$$\mathcal{M}\left\{\int_0^\infty \frac{1}{\xi}\phi\left(\frac{x}{\xi}\right)\psi(\xi)\,\mathrm{d}\xi\right\} = \Phi(s)\Psi(s), \qquad (3.159)$$

which is the **convolution** in the Mellin transform,[8] and conversely,

$$\mathcal{M}^{-1}\{\Phi(s)\,\Psi(s)\} = \int_0^\infty \frac{1}{\xi}\phi\left(\frac{x}{\xi}\right)\psi(\xi)\,\mathrm{d}\xi.$$

Generalization of the above results[9] may also be useful.

c. Further examples

Example 3.12. Before dealing with Example 3.13, we derive the relation

$$\mathcal{M}\{x^{-\nu}J_\nu(x)\} = \int_0^\infty x^{s-\nu-1}J_\nu(x)\,\mathrm{d}x$$

$$= \frac{2^{s-\nu-1}\Gamma\left(\dfrac{s}{2}\right)}{\Gamma\left(\nu - \dfrac{s}{2}+1\right)} \qquad \left(0 < s < \nu + \frac{3}{2}\right). \qquad (3.160)$$

---

[8]If we put $\mathrm{e}^u = \xi$, $\mathrm{e}^x = \eta$, $f(u) = f(\log\xi) = \psi(\xi) = \psi(\mathrm{e}^u)$, $g(x-u) = \phi(\mathrm{e}^{x-u}) = \phi(\eta/\xi)$ into the definition of the convolution in the Fourier transform Eq. (3.64), then we have

$$f * g = \int_{-\infty}^\infty f(u)g(x-u)\,\mathrm{d}u \quad \rightarrow \quad = \int_0^\infty \psi(\xi)\phi\left(\frac{\eta}{\xi}\right)\frac{\mathrm{d}\xi}{\xi}.$$

[9]Extension of Eq. (3.159) is given by

$$\mathcal{M}\left\{\int_0^\infty \cdots \int_0^\infty \frac{1}{\xi_1 \cdots \xi_n}\phi\left(\frac{x}{\xi_1 \cdots \xi_n}\right)\psi_1(\xi_1)\cdots\psi_n(\xi_n)\,\mathrm{d}\xi_1\cdots\mathrm{d}\xi_n\right\}$$
$$= \Phi(s)\,\Psi_1(s)\cdots\Psi_n(s),$$

where we denote $\Phi(s) = \mathcal{M}\{\phi\}$ and $\Psi_i(s) = \mathcal{M}\{\psi_i\}$ $(n = 1, \cdots n)$.

We first calculate the Fourier cosine transform of the function $f(x)$:

$$f(x) = \begin{cases} (1 - x^2)^{\nu - 1/2} & (0 \le x < 1) \\ 0 & (x > 1) \end{cases}.$$

Hereafter, we omit the normalization constant of the Fourier cosine transforms $\sqrt{2/\pi}$ unless otherwise stated. Consider the integral

$$I_1 \equiv \int_0^1 (1 - x^2)^{\nu - 1/2} \cos(kx) \, dx = \sum_{n=0}^{\infty} \frac{(-1)^n k^{2n}}{(2n)!} \int_0^1 x^{2n} (1 - x^2)^{\nu - 1/2} dx,$$

where we have expanded $\cos(kx)$ in the Taylor's series about $x = 0$. The integral over the interval $[0, 1]$ is rewritten by the well-known **beta function** by the change of variable $x^2 = t$ (see Appendix A.1.2):

$$\int_0^1 x^{2n}(1 - x^2)^{\nu - 1/2} dx = \frac{1}{2} \int_0^1 t^{n - 1/2}(1 - t)^{\nu - 1/2} dt$$

$$= \frac{1}{2} B\left(n + \frac{1}{2}, \nu + \frac{1}{2}\right) = \frac{1}{2} \frac{\Gamma\left(n + \frac{1}{2}\right) \Gamma\left(\nu + \frac{1}{2}\right)}{\Gamma(n + \nu + 1)}.$$

Then, it follows that

$$I_1 = \sum_{n=0}^{\infty} \frac{(-1)^n}{2(2n)!} \frac{\Gamma\left(n + \frac{1}{2}\right) \Gamma\left(\nu + \frac{1}{2}\right)}{\Gamma(n + \nu + 1)} k^{2n}$$

$$= \frac{\sqrt{\pi}}{2} \Gamma\left(\nu + \frac{1}{2}\right) \sum_{n=0}^{\infty} \frac{(-1)^n}{n! \Gamma(n + \nu + 1)} \left(\frac{k}{2}\right)^{2n}$$

$$= \frac{\sqrt{\pi}}{2} \Gamma\left(\nu + \frac{1}{2}\right) \left(\frac{k}{2}\right)^{-\nu} J_\nu(k),$$

owing to

$$\frac{\Gamma\left(n + \frac{1}{2}\right)}{(2n)!} = \frac{\left(n - \frac{1}{2}\right)\left(n - \frac{3}{2}\right)...\Gamma\left(\frac{1}{2}\right)}{(2n)(2n - 1)...3 \cdot 2 \cdot 1} = \frac{\sqrt{\pi}}{2^{2n} n!},$$

and the series expansion of the Bessel function $J_\nu(k)$ (see the power series expansion (A.13)). Consequently, the Fourier cosine transform of $f(x)$ becomes

$$\mathcal{F}_c\{f(x)\} = 2^{\nu - 1/2} \Gamma\left(\nu + \frac{1}{2}\right) k^{-\nu} J_\nu(k) \equiv F_c(k). \tag{3.161}$$

Next, we consider the Fourier cosine transform of x^{-s}. Using the result shown in 3.3.2a(i)-(3), where the parameter a is changed to $1 - s$, we obtain the Fourier cosine transform of $g(x) = x^{-s}$:

$$\mathcal{F}_c\{g(x)\} \equiv G_c(k) = \sqrt{\frac{2}{\pi}} \Gamma(1 - s) \cos\left(\frac{\pi(1 - s)}{2}\right) k^{s-1}, \quad (0 < s < 1).$$

Making use of the property of the gamma function $\Gamma(s)\Gamma(1-s) = \pi/\sin(\pi s)$ (see Appendix (A.4)), we may rewrite

$$G_c(k) = \sqrt{\frac{2}{\pi}} \frac{\pi \sin\left(\frac{\pi s}{2}\right)}{\Gamma(s) \sin(\pi s)} k^{s-1} = \sqrt{\frac{\pi}{2}} \frac{1}{\Gamma(s) \cos\left(\frac{\pi s}{2}\right)} k^{s-1}.$$

Using the Parseval's relation between $F_c(k)G_c(k)$ and $f(x)g(x)$,[10] we have

$$\frac{2^{\nu-1}\sqrt{\pi}\,\Gamma(\nu+\frac{1}{2})}{\Gamma(s) \cos\left(\frac{\pi s}{2}\right)} \int_0^\infty k^{s-\nu-1} J_\nu(k)\,dk = \int_0^1 (1-x^2)^{\nu-1/2} x^{-s}\,dx.$$

Furthermore, by putting $x^2 = t$, and rewriting the integral in terms of the beta function and the gamma function, we have

$$\text{r.h.s.} = \frac{1}{2} \int_0^1 (1-t)^{\nu-1/2} t^{-(s+1)/2}\,dt = \frac{1}{2} B\left(\nu+\frac{1}{2}, \frac{1}{2} - \frac{s}{2}\right)$$

$$= \frac{1}{2} \Gamma\left(\nu+\frac{1}{2}\right) \Gamma\left(\frac{1}{2} - \frac{s}{2}\right) / \Gamma\left(\nu - \frac{s}{2} + 1\right),$$

and hence we have

$$\int_0^\infty k^{s-\nu-1} J_\nu(k)\,dk = \frac{\Gamma\left(\frac{1}{2} - \frac{s}{2}\right) \Gamma(s) \cos\left(\frac{\pi s}{2}\right)}{2^\nu \sqrt{\pi} \Gamma\left(\nu - \frac{s}{2} + 1\right)}.$$

Further use of the property of the gamma function (Appendix (A.5), (A.6)) yields the result of Eq. (3.160):

$$\int_0^\infty k^{s-\nu-1} J_\nu(k)\,dk = \frac{2^{s-\nu-1} \Gamma\left(\frac{s}{2}\right)}{\Gamma\left(\nu - \frac{s}{2} + 1\right)}. \qquad (3.162)$$

◁

Example 3.13. An integral transform and its inverse integral, with the kernel $k(sx)$ and $h(sx)$, respectively,

$$\psi(s) = \int_0^\infty \phi(x)\,k(sx)\,dx, \quad \phi(x) = \int_0^\infty \psi(s)\,h(xs)\,ds \qquad (3.163)$$

---

[10]As is stated in the Fourier transforms,

$$\int_0^\infty F_c(k)\,\bar{G}_c(k)\,dk = \int_0^\infty f(x)\,\bar{g}(x)\,dx$$

holds in the Fourier cosine transforms, similar to Eq. (3.62).

are successfully performed only if the Mellin transforms of the respective kernels $K(s)(\equiv \mathcal{M}\{k\})$ and $H(s)(\equiv \mathcal{M}\{h\})$ have a relation

$$K(s)H(1-s) = 1 \quad \text{or} \quad K(1-s)H(s) = 1. \tag{3.164}$$

Indeed, if the relations (3.163) hold, the conditions (3.164) follow, which is shown as follows: Let $\Phi(s)$, $\Psi(s)$ be the Mellin transforms of $\phi(x)$, $\psi(x)$. Then,

$$
\begin{aligned}
\Phi(s) &= \int_0^\infty x^{s-1}\phi(x)\,\mathrm{d}x = \int_0^\infty x^{s-1}\left(\int_0^\infty \psi(t)\,h(xt)\,\mathrm{d}t\right)\mathrm{d}x \\
&= \int_0^\infty \psi(t)\,\mathrm{d}t \int_0^\infty x^{s-1}h(xt)\,\mathrm{d}x.
\end{aligned}
$$

By the change of variable from x to y $(= xt)$, the above equation becomes

$$r.h.s. = \int_0^\infty t^{-s}\psi(t)\,\mathrm{d}t \int_0^\infty y^{s-1}h(y)\,\mathrm{d}y = \Psi(1-s)H(s),$$

i.e., $\Phi(s) = \Psi(1-s)H(s)$. A similar relation $\Psi(s) = \Phi(1-s)K(s)$ is obtained by performing the Mellin transform to $\psi(x)$, so that we have a set of relations

$$\Phi(s) = \Psi(1-s)H(s), \quad \Psi(s) = \Phi(1-s)K(s).$$

From the first relation, we have $\Phi(1-s) = \Psi(s)H(1-s)$, which is substituted for the second relation, leading to

$$\Psi(s) = \Phi(1-s)K(s) = \Psi(s)H(1-s)K(s),$$

and hence $H(1-s)K(s) = 1$. By the symmetry of the transforms (or after a similar calculation), we obtain $K(1-s)H(s) = 1$. Although we skip the proof here, the relation (3.164) is the necessary and sufficient condition, in which the kernel of a given integral transform $K(x,s)$ and the kernel of its inverse integral transform $H(s,x)$ is described by a function of the product of s and x (*i.e.*, sx). ◁

(i) For example, consider $k(yx) = \sqrt{2/\pi}\sin(yx)$, where y is a given parameter.

From the present subsection §§3.3.4a(vii), we have

$$\mathcal{M}\{\sin(ax)\} = a^{-s}\Gamma(s)\sin\left(\frac{\pi}{2}s\right).$$

By putting $a = y = 1$ in the above equations, the Mellin transform of $k(x)$ is

$$K(s) = \mathcal{M}\left\{\sqrt{\frac{2}{\pi}}\sin x\right\} = \sqrt{\frac{2}{\pi}}\Gamma(s)\sin\left(\frac{\pi s}{2}\right).$$

Then, the second relation of Eq. (3.164) gives

$$H(s) = \frac{1}{K(1-s)} = \sqrt{\frac{\pi}{2}} \frac{1}{\Gamma(1-s)\cos\left(\frac{\pi}{2}s\right)} = \sqrt{\frac{2}{\pi}} \Gamma(s)\sin\left(\frac{\pi}{2}s\right),$$

where we make use of the property $\Gamma(z)\Gamma(1-z) = \pi/\sin(\pi z)$ (see Appendix (A.4)). The last equation is of the same form as the Mellin transform of $k(x)$, so that the kernel $h(x)$ for the inverse Mellin transform has the same form as $k(x)$.

(ii) If $k(x) = x^{1/2}J_\nu(x)$, the Mellin transform of $k(x)$ is

$$\mathcal{M}\{x^{1/2}J_\nu(x)\} = \frac{2^{s-1/2}\Gamma\left(\frac{\nu}{2} + \frac{s}{2} + \frac{1}{4}\right)}{\Gamma\left(\frac{\nu}{2} - \frac{s}{2} + \frac{3}{4}\right)} = K(s),$$

which is obtained by the use of Eq. (3.162) with $s \to \nu + s + 1/2$. Then, we obtain

$$H(s) = \frac{1}{K(1-s)} = \frac{2^{s-1/2}\Gamma\left(\frac{\nu}{2} + \frac{s}{2} + \frac{1}{4}\right)}{\Gamma\left(\frac{\nu}{2} - \frac{s}{2} + \frac{3}{4}\right)} = K(s),$$

and hence $h(x) = x^{1/2}J_\nu(x)$. These kernels provide the Hankel transforms described in the next subsection.

As shown in these examples, the kernel functions, in which the kernel of the integral transform $k(s,x)$ and that of the inverse integral transform $h(s,x)$ are of the same form, are called the **Fourier kernels**.

The Mellin transforms are appropriately applied to the boundary-value problems in a wedge-shaped boundary (which includes a region bounded by perpendicular walls, bounded by a single plane wall, and a region outside a semi-infinite plate, as a special case). We show such an application.

Example 3.14 (Steady temperature distribution inside a wedge-shaped region).

We consider a steady temperature distribution bounded by two infinitely large intersecting planes (inside an infinitely large wedge-shaped region) with intersecting angle 2α, which is assumed here to be less than π. This is a two-dimensional problem, and the choice of the polar coordinate system (r, θ) with its origin at the intersection seems appropriate. In the latter, the boundaries are specified by $\theta = -\alpha$ and α. The basic time-dependent equation will be the diffusion equation or heat conduction equation (2.5) developed in §§2.1.2, where the number of spatial dimensions is extended to two. In the present steady state problem, however, it falls into a two-dimensional Laplace equation:

$$r\frac{\partial}{\partial r}\left(r\frac{\partial T}{\partial r}\right) + \frac{\partial^2 T}{\partial \theta^2} = 0, \tag{3.165}$$

which is of the form of ordinary Laplace equation multiplied by r^2.

In the following, we consider the Dirichlet-type boundary condition:

$$T = f(r) \quad \text{at} \quad \theta = -\alpha \; ; \quad T = g(r) \quad \text{at} \quad \theta = \alpha, \quad (f(0) = g(0)),$$

$$T \to 0 \quad \text{for} \quad 0 \leq |\theta| < \alpha \quad \text{and} \quad r \to \infty. \tag{3.166}$$

By applying the Mellin transform:

$$\mathcal{M}\{T(r,\theta)\} = \int_0^\infty T(r,\theta)\, r^{s-1} \mathrm{d}r \equiv \Phi(s,\theta)$$

to Eq. (3.165), and by using §§3.3.4b(v), we have

$$\left(\frac{\mathrm{d}^2}{\mathrm{d}\theta^2} + s^2\right)\Phi(s,\theta) = 0, \tag{3.167}$$

from which the general solution

$$\Phi(s,\theta) = A(s)\cos(s\theta) + B(s)\sin(s\theta) \tag{3.168}$$

is obtained. Boundary conditions (3.166) are also transformed:

$$\Phi(s,-\alpha) = \mathcal{M}\{f(r)\} = F(s), \quad \Phi(s,\alpha) = \mathcal{M}\{g(r)\} = G(s), \tag{3.169}$$

and A, B are determined to satisfy Eqs. (3.169). Accordingly, we obtain the solution of Eq. (3.165) in the transformed space:

$$\Phi(s,\theta) = \frac{F(s)\sin[(\alpha - \theta)s] + G(s)\sin[(\alpha + \theta)s]}{\sin(2\alpha s)}. \tag{3.170}$$

By applying the inverse Mellin transform to Eq. (3.170),

$$T(r,\theta) = \mathcal{M}^{-1}\{\Phi(s,\theta)\} = \frac{1}{2\pi\mathrm{i}}\int_{\gamma-\mathrm{i}\infty}^{\gamma+\mathrm{i}\infty} \Phi(s,\theta) r^{-s}\mathrm{d}s, \tag{3.171}$$

we can obtain the solution in the original physical space.

As an example, if the boundary conditions are symmetric (and hence $G = F$), Eq. (3.170) becomes

$$\Phi(s,\theta) = \frac{F(s)\cos(\theta s)}{\cos(\alpha s)}. \tag{3.172}$$

Furthermore, if $f = g = 1$ for $0 \le x \le 1$, and $f = g = 0$ for $x > 1$, we have

$$T(r,\theta) = \mathcal{M}^{-1}\left\{\frac{\cos(\theta s)}{s\cos(\alpha s)}\right\} = \frac{1}{2\pi i}\int_{\gamma-i\infty}^{\gamma+i\infty}\frac{\cos(\theta s)}{s\cos(\alpha s)}r^{-s}ds. \quad (3.173)$$

To perform the latter integral, we consider the contour integral in the complex s plane, as shown in Fig. 3.10(a). Note that the singularities of the integrand, $s = 0$, $(n+1/2)\pi/\alpha$, $n = 0, \pm1, \pm2, \cdots$, are all simple poles, and that the pole $\pi/2\alpha$ nearest to the imaginary axis is larger than 1 for $2\alpha < 180°$, so that $0 < \gamma < 1 < \pi/2\alpha$. We also need to choose the contour depending on $0 < r < 1$ or $r > 1$, which determines the behavior of $r^{-s} = \exp(-s\log r)$.

(1) Case $0 < r < 1$

In this case, we choose the contour C_0 parallel to the imaginary axis $s = \gamma - iR \to \gamma + iR$, and the semi-circle C_L of radius R in the left-half plane (plus line segments at $\operatorname{Im} s = \pm R$) to form a closed contour in the anti-clockwise direction, as shown in Fig. 3.10(a). By summing up all residues inside the closed contour, and taking the limit $R \to \infty$, we obtain

$$T = \frac{1}{2\pi i}\int_{C_0}\cdots ds = \frac{1}{2\pi i}\int_{C_0+C_L}\cdots ds$$

$$= \operatorname{Res}(s=0) + \sum_{n=0}^{\infty}\operatorname{Res}\left(s = -\frac{(2n+1)\pi}{2\alpha}\right)$$

$$= 1 - \sum_{n=0}^{\infty}\frac{2(-1)^n}{(2n+1)\pi}r^{(2n+1)\pi/2\alpha}\cos\left(\frac{(2n+1)\pi}{2\alpha}\theta\right). \quad (3.174)$$

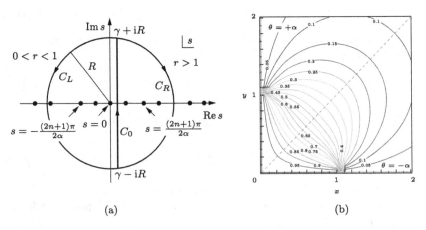

Fig. 3.10 (a) Contour integral, and (b) a steady temperature distribution in an infinite wedge.

(2) Case $r > 1$

In this case, we choose the contour C_0 and the semi-circle C_R of radius R in the right-half plane to form a closed contour in the clockwise direction, as shown in Fig. 3.10(a). By summing up all residues inside the closed contour, and taking the limit $R \to \infty$, we obtain

$$T = \sum_{n=0}^{\infty} \frac{2(-1)^n}{(2n+1)\pi} r^{-(2n+1)\pi/2\alpha} \cos\left(\frac{(2n+1)\pi}{2\alpha}\theta\right). \qquad (3.175)$$

We show an example of the temperature distribution in Fig. 3.10(b), where $\alpha = \pi/4$, and the boundary planes $\theta = -\alpha$ and $\theta = +\alpha$ are chosen along the x axis and the y axis, respectively. ◁

3.3.5 *Hankel transforms*

The double Fourier transform to a given function $f(x, y)$ of the two independent variables x, y defined in $-\infty < x < \infty, -\infty < y < \infty$ is

$$F(\xi, \eta) = \frac{1}{2\pi} \int_{-\infty}^{\infty} \int_{-\infty}^{\infty} e^{-i(\xi x + \eta y)} f(x, y) \, dx \, dy, \qquad (3.176)$$

where we assume the existence of the integral. The inverse transform is then given by

$$f(x, y) = \frac{1}{2\pi} \int_{-\infty}^{\infty} \int_{-\infty}^{\infty} e^{i(\xi x + \eta y)} F(\xi, \eta) \, d\xi \, d\eta. \qquad (3.177)$$

We now change the variable (x, y) in the rectangular coordinate system to (r, θ) of the polar coordinate system. We also introduce ρ, ϕ in the polar coordinate system corresponding to ξ, η, so that $\xi x + \eta y = \rho r \cos(\phi - \theta)$. Then, Eq. (3.176) becomes

$$F(\rho, \phi) = \frac{1}{2\pi} \int_0^{\infty} dr\, r \int_0^{2\pi} d\theta\, e^{-i\rho r \cos(\phi - \theta)} f(r, \theta). \qquad (3.178)$$

If we confine our attention to an axisymmetric case, f depends only on r, and the angular variables θ, ϕ have a meaning only through the relative angle $\phi - \theta$. Introducing a new angular variable $\varphi = \phi - \theta - \pi/2$, we have

$$F(\rho) = \frac{1}{2\pi} \int_0^{\infty} dr\, r \int_0^{2\pi} d\theta\, e^{-i\rho r \cos(\phi - \theta)} f(r)$$
$$= \int_0^{\infty} dr\, r f(r) \left(\frac{1}{2\pi} \int_{\beta}^{2\pi + \beta} e^{i\rho r \sin \varphi} \, d\varphi \right),$$

where β is an arbitrary constant with a view to axisymmetry. Furthermore, the use of the integral representation of the Bessel function (A.25)

$$J_n(x) = \frac{1}{2\pi} \int_\alpha^{2\pi+\alpha} e^{i(x\sin\theta - n\theta)} d\theta, \qquad (3.179)$$

with $n = 0$ leads to

$$F(\rho) = \int_0^\infty f(r) r J_0(\rho r)\, dr, \qquad (3.180)$$

which is the Hankel transform (3.51) with $n = 0$. The inverse Hankel transform of (3.180) are derived similarly as

$$f(r) = \int_0^\infty F(\rho)\, \rho J_0(\rho r)\, d\rho. \qquad (3.181)$$

◁

The Hankel transforms are generally defined by Eq. (3.51) or Eq. (3.52) using the ν-th order Bessel function:

$$\mathcal{H}\{u(x)\} \equiv \int_0^\infty u(x)\, x J_\nu(\xi x)\, dx = U(\xi),$$

$$u(x) = \int_0^\infty U(\xi)\, \xi J_\nu(x\xi)\, d\xi = \mathcal{H}^{-1}\{U(\xi)\}.$$

In the special case $\nu = 1/2$, the Bessel function is reduced to $J_{1/2}(x) = \sqrt{2/(\pi x)}\, \sin x$ (see Eq. (A.19)), so that we have

$$U(\xi) = \sqrt{\frac{2}{\pi\xi}} \int_0^\infty \sqrt{x}\, u(x) \sin(\xi x)\, dx.$$

By introducing new functions $v(x) = \sqrt{x}\, u(x)$ and $V(\xi) = \sqrt{\xi}\, U(\xi)$, the above relations become

$$V(\xi) = \sqrt{\frac{2}{\pi}} \int_0^\infty v(x) \sin(\xi x)\, dx, \quad v(x) = \sqrt{\frac{2}{\pi}} \int_0^\infty V(\xi) \sin(\xi x)\, d\xi, \qquad (3.182)$$

which are the Fourier sine transforms. Similarly, taking into account that $J_{-1/2}(x) = \sqrt{2/(\pi x)}\, \cos x$ (the case of $\nu = -1/2$), the above relations are reduced to the Fourier cosine transforms.

a. Simple examples

(i) The Fourier–Bessel integral formula (3.53) for $0 \le x < \infty$, $0 \le \xi < \infty$ is regarded as

$$\delta(x - \xi) = \xi \int_0^\infty \rho J_n(x\rho) J_n(\xi\rho)\, d\rho,$$

which is also regarded as the Hankel transform of $J_n(\lambda\rho)$, ($\lambda = x$, or ξ), with the use of the Bessel function of n-th order.

In the following, we consider the Hankel transforms with $\nu = 0$, unless otherwise stated,

(ii) The integrals over the interval $[0, \infty)$ whose integrands include the Bessel function are regarded as examples of the Hankel transforms. The example (3.102), shown as an example of the Laplace transform, is symmetric with s and k, so that

$$\frac{1}{\sqrt{s^2 + k^2}} = \int_0^\infty e^{-sx} J_0(kx) \, dx = \int_0^\infty e^{-kx} J_0(sx) \, dx.$$

The latter can be regarded as the Hankel transform of e^{-kx}/x. Furthermore, if we put $k = i\kappa$ in the above equation, we obtain

$$\frac{1}{\sqrt{s^2 - \kappa^2}} = \int_0^\infty e^{-i\kappa x} J_0(sx) \, dx = \int_0^\infty [\cos(\kappa x) - i\sin(\kappa x)] J_0(sx) \, dx,$$

from which we have

$$\int_0^\infty J_0(sx) \cos(\kappa x) \, dx = \begin{cases} 0 & (|s| < |\kappa|) \\ \dfrac{1}{\sqrt{s^2 - \kappa^2}} & (|s| > |\kappa|) \end{cases}, \qquad (3.183)$$

$$\int_0^\infty J_0(sx) \sin(\kappa x) \, dx = \begin{cases} \dfrac{1}{\sqrt{\kappa^2 - s^2}} & (|s| < |\kappa|) \\ 0 & (|s| > |\kappa|) \end{cases}. \qquad (3.184)$$

These expressions may be interpreted as the Fourier cosine transform and the Fourier sine transform of $J_0(sx)$, respectively (except for the normalization constants), or may be interpreted as the Hankel transforms of $\cos(\kappa x)/x$ and $\sin(\kappa x)/x$, respectively.

(iii) By integrating Eq. (3.183) with respect to κ in the interval $[0, \kappa]$, we have

$$\int_0^\infty J_0(sx) \frac{\sin(\kappa x)}{x} \, dx = \begin{cases} \pi/2 & (|s| \le |\kappa|) \\ \arcsin(\kappa/s) & (|s| \ge |\kappa|) \end{cases}, \qquad (3.185)$$

which may be interpreted as the Fourier sine transform of $J_0(sx)/x$ (except for the normalization constant), or may be interpreted as the Hankel transform of $\sin(\kappa x)/x^2$.

(iv) Furthermore, if the previous relation Eq. (3.185) is differentiated with respect to s, we have

$$\int_0^\infty J_1(sx) \sin(\kappa x) \, dx = \begin{cases} 0 & (|s| < |\kappa|) \\ \dfrac{\kappa}{s\sqrt{s^2 - \kappa^2}} & (|s| > |\kappa|) \end{cases}, \qquad (3.186)$$

which may be interpreted as the Fourier sine transform of $J_1(sx)$ (except for the normalization constant), or may be interpreted as the Hankel transform of $\sin(\kappa x)/x$ with J_1.

(v) If we put $\nu = 0$ in the relation (3.162) that are derived as an example of the Mellin transform, we have

$$\int_0^\infty k^{s-1} J_0(k)\, dk = \frac{2^{s-1} \Gamma\left(\dfrac{s}{2}\right)}{\Gamma\left(1 - \dfrac{s}{2}\right)}.$$

By changing the variable k to r, such that $k = r\rho$, we have

$$\rho^s \int_0^\infty r^{s-1} J_0(r\rho)\, dr = \frac{2^{s-1} \Gamma\left(\dfrac{s}{2}\right)}{\Gamma\left(1 - \dfrac{s}{2}\right)} = \frac{2^{s-1}}{\pi} \Gamma\left(\dfrac{s}{2}\right)^2 \sin\left(\dfrac{\pi s}{2}\right),$$

or

$$\mathcal{H}\{r^{s-2}\} = \frac{2^{s-1}}{\pi \rho^s} \Gamma\left(\frac{s}{2}\right)^2 \sin\left(\frac{\pi s}{2}\right), \tag{3.187}$$

where Eq. (A.4) is used. In particular, the substitution of $s = 1$ in the above equation yields

$$\mathcal{H}\left\{\frac{1}{r}\right\} = \int_0^\infty J_0(\rho r)\, dr = \frac{1}{\rho}. \tag{3.188}$$

b. General formulae of the Hankel transforms

Hankel transforms appear when the Fourier transforms are given in the polar coordinate system, and are suitable for problems with an axisymmetric boundary. However, there are not as many simple general formulae as are developed in the Fourier transforms. We demonstrate a few of them.

(i) Multiplication of the variable by a constant

$$\mathcal{H}\{f(ar)\} = \int_0^\infty r f(ar)\, J_0(\rho r)\, dr$$

$$= \frac{1}{a^2} \int_0^\infty \xi f(\xi)\, J_0\left(\frac{\rho}{a}\xi\right) d\xi = \frac{1}{a^2} F\left(\frac{\rho}{a}\right).$$

(ii) Parseval's relation

$$\int_0^\infty \rho F(\rho) G(\rho)\, d\rho = \int_0^\infty \rho\, G(\rho)\, d\rho \int_0^\infty r f(r)\, J_0(\rho r)\, dr$$

$$= \int_0^\infty r\, f(r)\, dr \int_0^\infty \rho G(\rho)\, J_0(\rho r)\, d\rho$$

$$= \int_0^\infty r f(r)\, g(r)\, dr, \tag{3.189}$$

which are called the **Parseval's relation**.

For further detailed formulae of the Hankel transforms, including those which use $J_\nu(\rho r)$, see *e.g.*, [Erdélyi (1954)](Chap. VIII),

c. Further examples

Example 3.15 (Green's function of modified Helmholtz equation). We shall obtain the principal part of the Green's function of the two-dimensional modified Helmholtz equation:

$$(\Delta - k^2)g = \frac{\partial^2 g}{\partial x^2} + \frac{\partial^2 g}{\partial y^2} - k^2 g = -\delta(x - \xi)\,\delta(y - \eta). \qquad (3.190)$$

If we choose the origin of the polar coordinate system (r, θ) at the singular point, and take account of the symmetry, Eq. (3.190) becomes

$$\frac{\mathrm{d}^2 g}{\mathrm{d}r^2} + \frac{1}{r}\frac{\mathrm{d}g}{\mathrm{d}r} - k^2 g = -\frac{1}{2\pi r}\delta(r), \qquad (3.191)$$

to which the Hankel transform $\mathcal{H}\{g(r)\} \equiv G(\rho)$ is applied. Noting the following calculation

$$
\begin{aligned}
\mathcal{H}\left\{\frac{\mathrm{d}^2 g}{\mathrm{d}r^2} + \frac{1}{r}\frac{\mathrm{d}g}{\mathrm{d}r}\right\} &= \mathcal{H}\left\{\frac{1}{r}\frac{\mathrm{d}}{\mathrm{d}r}\left(r\frac{\mathrm{d}g}{\mathrm{d}r}\right)\right\} = \int_0^\infty \frac{\mathrm{d}}{\mathrm{d}r}\left(r\frac{\mathrm{d}g}{\mathrm{d}r}\right) J_0(\rho r)\,\mathrm{d}r \\
&= \left[r\frac{\mathrm{d}g}{\mathrm{d}r}J_0(\rho r)\right]_0^\infty - \int_0^\infty r\frac{\mathrm{d}g}{\mathrm{d}r}J_0'(\rho r)\rho\,\mathrm{d}r \\
&= -\rho\int_0^\infty g'(r)\,rJ_0'(\rho r)\,\mathrm{d}r \\
&= \rho\int_0^\infty g(r)\left[\rho r J_0''(\rho r) + J_0'(\rho r)\right]\,\mathrm{d}r \\
&= -\rho^2\int_0^\infty r\,g(r)\,J_0(\rho r)\,\mathrm{d}r = -\rho^2 G(\rho),
\end{aligned}
$$

where the integration by parts and the equation satisfied by J_0 are made use of, we obtain

$$G(\rho) = \frac{1}{2\pi}\frac{1}{\rho^2 + k^2}. \qquad (3.192)$$

By applying the inverse Hankel transform, we obtain

$$g(r) = \frac{1}{2\pi}\int_0^\infty \frac{\rho}{\rho^2 + k^2}J_0(r\rho)\,\mathrm{d}\rho = \frac{1}{2\pi}K_0(kr), \qquad (3.193)$$

where $K_0(x)$ is the modified Bessel function of the second kind (see Appendix A.2.4). To obtain the last expression, see *e.g.*, [Abramowitz and Stegun (1964)] formula (11.4.44). The present result is the Green's function given in Table 2.2, column (c), two-dimensional case. ◁

Example 3.16 (Potential around a disk). We consider the electrostatic potential ϕ around a disk of radius a which is charged to a prescribed potential ϕ_0. We choose the cylindrical coordinate system (ρ, φ, z), such that the z axis is chosen perpendicular to the disk, and that the origin is taken at the center of the disk. Then the potential becomes axisymmetric (independent of φ), and should be a function of the distance z from the plane of the disk and the distance ρ from the z axis.

The governing equation for ϕ is

$$\Delta\phi(\rho, z) = \frac{\partial^2\phi}{\partial\rho^2} + \frac{1}{\rho}\frac{\partial\phi}{\partial\rho} + \frac{\partial^2\phi}{\partial z^2} = \frac{1}{\rho}\frac{\partial}{\partial\rho}\left(\rho\frac{\partial\phi}{\partial\rho}\right) + \frac{\partial^2\phi}{\partial z^2} = 0, \qquad (3.194)$$

and the boundary conditions are

$$\begin{cases} \phi = \phi_0 & (0 \le \rho \le a) \\ \dfrac{\partial\phi}{\partial z} = 0 & (\rho > a) \end{cases} \qquad \text{at} \quad z = 0, \qquad (3.195)$$

$$\phi \to 0 \qquad \text{as} \quad |z| \to \infty.$$

The problems with boundary conditions such as (3.195), in which the value of ϕ is given at some part of the boundary, whereas its derivative is given at the rest of the boundary, are called **mixed boundary-value problems**.

Taking account of the symmetry with respect to z, we consider $z \ge 0$ hereafter. By applying the Hankel transform to Eq. (3.194):

$$\mathcal{H}\{\phi(\rho, z)\} = \int_0^\infty \rho\,\phi(\rho, z)J_0(k\rho)\,\mathrm{d}\rho \equiv \Phi(k, z), \qquad (3.196)$$

and by performing the integration with respect to ρ, we obtain

$$\begin{aligned} \frac{\partial^2\Phi}{\partial z^2} &= \int_0^\infty \rho\frac{\partial^2\phi}{\partial z^2}J_0(k\rho)\,\mathrm{d}\rho = -\int_0^\infty \frac{\partial}{\partial\rho}\left(\rho\frac{\partial\phi}{\partial\rho}\right)J_0(k\rho)\,\mathrm{d}\rho \\ &= \left[-\rho\frac{\partial\phi}{\partial\rho}J_0(k\rho)\right]_0^\infty + k\int_0^\infty \rho\frac{\partial\phi}{\partial\rho}J_0'(k\rho)\,\mathrm{d}\rho \\ &= [k\rho\phi J_0'(k\rho)]_0^\infty - k\int_0^\infty \phi\left(k\rho J_0''(k\rho) + J_0'(k\rho)\right)\,\mathrm{d}\rho \\ &= k^2\int_0^\infty \phi\rho J_0(k\rho)\,\mathrm{d}\rho = k^2\Phi. \end{aligned} \qquad (3.197)$$

The basic solutions of Eq. (3.197) are $\exp(\pm kz)$, from which we choose the one in which $\phi \to 0$ as $z \to \infty$. Accordingly, we obtain the solution $\Phi(k, z) = A(k)\exp(-kz)$ in the transformed space, where $A(k)$ is an arbitrary constant to be determined by the boundary conditions. We now apply the inverse Hankel transform,

$$\phi(\rho, z) = \int_0^\infty k\,A(k)\mathrm{e}^{-kz}J_0(k\rho)\,\mathrm{d}k, \qquad (3.198)$$

and impose the boundary condition (3.195):

$$\int_0^\infty k A(k)\, J_0(k\rho)\, \mathrm{d}k = \phi_0 \qquad (0 \le \rho \le a), \qquad (3.199)$$

$$\int_0^\infty k^2 A(k)\, J_0(k\rho)\, \mathrm{d}k = 0 \qquad (\rho > a). \qquad (3.200)$$

As stated, the boundary value ϕ is given only at $0 \le \rho \le a$, so that we assume $\phi = g(\rho)$ at $\rho > a$, and seek the unknown functions $g(\rho)$ and $A(k)$ that satisfy the boundary conditions (3.199) and (3.200). Such a set of equations for determining the function $A(k)$ is called the "dual integral equations" (for more details, see [Sneddon (1966)]).

To do so, we remark the following relation to find a candidate of the solution, because it satisfies one of the boundary conditions (3.199):

$$\int_0^\infty \frac{\sin(ak)}{k} J_0(k\rho)\, \mathrm{d}k = \begin{cases} \dfrac{\pi}{2} & (0 \le \rho \le a) \\[2mm] \arcsin\left(\dfrac{a}{\rho}\right) & (\rho \ge a) \end{cases}, \qquad (3.201)$$

where we use Eq. (3.185) with $\kappa = a$, $s = \rho$ and $x = k$. Comparing the above equation with the boundary condition (3.199), we have

$$A(k) = \frac{2\phi_0}{\pi} \frac{\sin(ak)}{k^2}, \qquad (3.202)$$

(*cf.* [Sneddon (1966)](§3.1)). The substitution of $A(k)$ for the boundary condition (3.200), with the use of Eq. (3.184), yields

$$\frac{2\phi_0}{\pi} \int_0^\infty \sin(ak)\, J_0(k\rho)\, \mathrm{d}k = \begin{cases} \dfrac{2\phi_0}{\pi} \dfrac{1}{\sqrt{a^2 - \rho^2}} & (\rho < a) \\[2mm] 0 & (\rho > a) \end{cases},$$

which satisfies the remaining boundary condition for $\rho > a$.

By substituting Eq. (3.202) for Eq. (3.198), we have

$$\phi(\rho, z) = \frac{2\phi_0}{\pi} \int_0^\infty \frac{\sin(ak)}{k} \mathrm{e}^{-kz} J_0(k\rho)\, \mathrm{d}k, \qquad (3.203)$$

and by performing the integral of the r.h.s., we obtain our final result:

$$\phi(\rho, z) = \frac{2\phi_0}{\pi} \arcsin\left(\frac{2a}{\sqrt{z^2 + (a+\rho)^2} + \sqrt{z^2 + (a-\rho)^2}} \right), \qquad (3.204)$$

(*cf.* [Titchmarsh (1948)](§11.16)).

We show the equi-potential curves around the disk in Fig. 3.11(a). At a far distant position on the axis of the disk $\rho = 0$, $z \gg a$, ϕ is

$$\phi(0, z) = \frac{2\phi_0}{\pi} \arcsin\left(\frac{a}{\sqrt{z^2 + a^2}} \right) \sim \frac{2\phi_0}{\pi} \frac{a}{z},$$

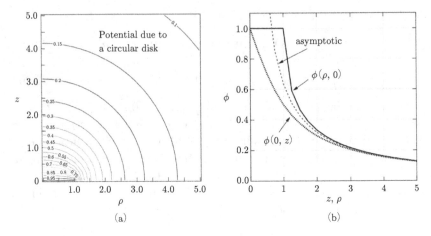

Fig. 3.11   (a) Equi-potential curves around a disk ($\phi = 1$ on the disk of radius $a = 1$). (b) Potential on the axis of the disk $\phi(0, z)$ and on the plane of the disk $\phi(\rho, 0)$.

whereas at a far distant position on the plane of the disk $z = 0$, $\rho \gg a$,

$$\phi(\rho, 0) \sim \frac{2\phi_0}{\pi} \frac{a}{\rho}.$$

Both cases show that the potential at sufficiently distant positions decreases inversely proportional to the distance from the disk (Fig. 3.11(b)).   ◁

3.3.6   *Application of integral transforms to integral equations*

Many problems in science and engineering are described by the differential equations. In problems such as certain types of diffusion, diffraction, and transport, however, we need to know the details of the origin (cause) based on the observed data (effect). The latter is often given in the form of integrals, in which the unknown function is included. This type of equation is called an **integral equation**. An integral equation is classified into two types: the **Fredholm equation**, in which the interval of the integral is fixed, and the **Volterra equation**, in which one end of the interval of the integral is variable. Furthermore, they are also classified based on whether the unknown function appears only in the integrand (the first kind), or appears both in and outside of the integrand (the second kind).

Consequently, the equation for the unknown function $\phi(t)$ is determined under the given known functions $f(x)$ and $k(x, t)$ (called a kernel):

(1) Fredholm equation of the first kind

$$f(x) = \int_a^b k(x,t)\,\phi(t)\,\mathrm{d}t, \qquad (3.205)$$

(2) Fredholm equation of the second kind

$$\phi(x) = f(x) + \lambda \int_a^b k(x,t)\,\phi(t)\,\mathrm{d}t, \qquad (3.206)$$

(3) Volterra equation of the first kind

$$f(x) = \int_a^x k(x,t)\,\phi(t)\,\mathrm{d}t, \qquad (3.207)$$

(4) Volterra equation of the second kind

$$\phi(x) = f(x) + \lambda \int_a^x k(x,t)\,\phi(t)\,\mathrm{d}t. \qquad (3.208)$$

Example 3.17. The integral transforms dealt with thus far may be regarded as integral equations. For example, the Fourier transform

$$f(x) = \frac{1}{\sqrt{2\pi}} \int_{-\infty}^{\infty} \mathrm{e}^{-\mathrm{i}xt}\phi(t)\,\mathrm{d}t$$

may be regarded as the Fredholm equation of the first kind for obtaining unknown $\phi(t)$, under the given function $f(x)$ and kernel $\mathrm{e}^{-\mathrm{i}xt}/\sqrt{2\pi}$. The solution is

$$\phi(t) = \frac{1}{\sqrt{2\pi}} \int_{-\infty}^{\infty} \mathrm{e}^{\mathrm{i}tx} f(x)\,\mathrm{d}x.$$

Similarly, the Laplace, Mellin and Hankel transforms are regarded as the Fredholm equations of the first kind with the interval of integral $[0, \infty)$, and the kernels e^{-xt}, t^{x-1} and $tJ_n(xt)$, respectively. Respective solutions are given by the corresponding inverse integral transforms. ◁

Numerous methods for solving integral equations are known, *e.g.*, [Titchmarsh (1948)](Chap. 11), [Morse and Feshbach (1953)](Chap. 8), [Arfken (1985)](Chap. 16), *etc.* However, we shall confine our attention to the methods, in which integral transforms have been successfully applied. In particular, we show some examples in which the integral equations are linear (unless otherwise stated) and the kernels $k(x,t)$ are of the form $k(x-t)$ as is often encountered in practice.

a. Case 1: Interval of integration $(-\infty, \infty)$

(i) Fredholm equation of the first kind

The Fredholm equation of the first kind (3.205) with the above-mentioned kernel is written in terms of the Fourier-type convolution (3.64):

$$f(x) = \int_{-\infty}^{\infty} k(x - t)\, \phi(t)\, dt = k * \phi.$$

By applying the Fourier transform, and using its convolution theorem, we have (see Eq. (3.65))

$$F(\omega) = \sqrt{2\pi} K(\omega)\, \Phi(\omega),$$

where we denote the Fourier transforms as $F(\omega) = \mathcal{F}\{f(x)\}$, $K(\omega) = \mathcal{F}\{k(x)\}$, and $\Phi(\omega) = \mathcal{F}\{\phi(x)\}$. The solution in the transformed space is given by

$$\Phi(\omega) = \frac{F(\omega)}{\sqrt{2\pi} K(\omega)},$$

from which the solution in the original space is obtained:

$$\phi(x) = \frac{1}{2\pi} \int_{-\infty}^{\infty} \frac{F(\omega)}{K(\omega)}\, e^{ix\omega}\, d\omega. \tag{3.209}$$

(ii) Fredholm equation of the second kind

Similarly, the Fredholm equation of the second kind (3.206) is described as

$$\phi(x) = f(x) + \lambda \int_{-\infty}^{\infty} k(x - t)\, \phi(t)\, dt = f + \lambda k * \phi.$$

By applying the Fourier transform, we have

$$\Phi(\omega) = F(\omega) + \sqrt{2\pi} \lambda K(\omega)\, \Phi(\omega),$$

from which we have[11]

$$\Phi(\omega) = \frac{F(\omega)}{1 - \sqrt{2\pi} \lambda K(\omega)},$$

---

[11]Note that the present method of solution holds for some types of nonlinear integral equations. For example, if the latter has the following form

$$\phi(x) = f(x) + \lambda \int_{-\infty}^{\infty} \phi(x - t)\, \phi(t)\, dt = f + \lambda \phi * \phi,$$

the convolution theorem is also applicable. Indeed, by applying the Fourier transform to the above equation, we have

$$\Phi = F + \sqrt{2\pi} \lambda \Phi^2,$$

from which we have

$$\Phi = \frac{1}{2\sqrt{2\pi}\lambda} \left(1 \pm \sqrt{1 - 4\sqrt{2\pi}\lambda F} \right).$$

By applying the inverse Fourier transform, we obtain the unknown function ϕ. This situation is the same in the nonlinear Volterra equation of the second kind (next item b) if the kernel is given by ϕ with delayed variable type, such as $x - t$ versus t.

and the solution in the original space

$$\phi(x) = \frac{1}{\sqrt{2\pi}} \int_{-\infty}^{\infty} \frac{F(\omega)}{1 - \sqrt{2\pi}\lambda K(\omega)} \, e^{ix\omega} \, d\omega. \qquad (3.210)$$

b. Case 2: Interval of integration $[0, x]$
(i) Volterra equation of the first kind

In this case, the Volterra equation (3.207) is described by using the Laplace-type convolution defined by Eq. (3.116) (§§3.3.3, b-(vi)):

$$f(x) = \int_0^x k(x - t) \, \phi(t) \, dt = k * \phi.$$

By applying the Laplace transform, and using its convolution theorem, we have (see Eq. (3.117))

$$F(s) = K(s) \, \Phi(s),$$

where we denote the Laplace transforms as $F(s) = \mathcal{L}\{f(x)\}$, $K(s) = \mathcal{L}\{k(x)\}$, and $\Phi(s) = \mathcal{L}\{\phi(x)\}$. The solution in the transformed space is given by

$$\Phi(s) = \frac{F(s)}{K(s)},$$

from which the solution in the original space is obtained:

$$\phi(x) = \frac{1}{2\pi i} \int_{\gamma-i\infty}^{\gamma+i\infty} \frac{F(s)}{K(s)} \, e^{xs} \, ds. \qquad (3.211)$$

(ii) Volterra equation of the second kind

Similarly, the Volterra equation of the second kind (3.208) is described as

$$\phi(x) = f(x) + \lambda \int_0^x k(x - t) \, \phi(t) \, dt = f + \lambda k * \phi.$$

By applying the Laplace transform, we have

$$\Phi(s) = F(s) + \lambda K(s) \, \Phi(s),$$

from which we have

$$\Phi(s) = \frac{F(s)}{1 - \lambda K(s)},$$

and the solution in the original space

$$\phi(x) = \frac{1}{2\pi i} \int_{\gamma-i\infty}^{\gamma+i\infty} \frac{F(s)}{1 - \lambda K(s)} \, e^{xs} \, ds. \qquad (3.212)$$

Example 3.18. Solve the **Abel's equation**

$$f(x) = \int_0^x \frac{\phi(t)}{(x-t)^\alpha}\, dt, \qquad (0 < \alpha < 1, \quad x > 0), \qquad (3.213)$$

where $\phi(x)$ is the unknown function, and $f(x)$ is a known function. This is the Volterra equation of the first kind, so that we can solve it by means of the Laplace transforms as well as the convolution theorem:

$$\mathcal{L}\{f(x)\} = F(s), \quad \mathcal{L}\{\phi(x)\} = \Phi(s), \quad \mathcal{L}\{x^{-\alpha}\} = \frac{(-\alpha)!}{s^{1-\alpha}}.$$

The solution in the transformed space is then given by

$$\Phi(s) = \frac{s^{1-\alpha}}{(-\alpha)!} F(s).$$

Rewriting $(-\alpha)!$ using the property of the gamma function Eq. (A.4), we obtain

$$\Phi(s) = \frac{\sin(\pi\alpha)}{\pi} \frac{(\alpha-1)! F(s)}{s^{\alpha-1}} = \frac{\sin(\pi\alpha)}{\pi} s\mathcal{L}\{x^{\alpha-1}\} F(s),$$

or

$$\frac{1}{s}\Phi(s) = \frac{\sin(\pi\alpha)}{\pi} \mathcal{L}\{x^{\alpha-1}\} F(s).$$

By applying the inverse Laplace transform, we obtain the solution $\phi(x)$. To do this, we note that the l.h.s. is the Laplace transform of the integral of ϕ (by the formula §3.3.3b(v), Eq. (3.115)), whereas the r.h.s. is given in the convolution form (§§3.3.3b(vi), Eq. (3.117)), so that we obtain

$$\phi(x) = \frac{\sin(\pi\alpha)}{\pi} \frac{d}{dx} \int_0^x \frac{f(t)}{(x-t)^{1-\alpha}}\, dt. \qquad (3.214)$$

◁

The solution of a slightly extended integral equation of the Abel-type:

$$f(x) = \int_a^x \frac{\phi(t)}{(x^2 - t^2)^\alpha}\, dt, \qquad (0 < \alpha < 1, \quad a < x) \qquad (3.215)$$

is also given by

$$\phi(x) = \frac{2\sin(\pi\alpha)}{\pi} \frac{d}{dx} \int_a^x \frac{t\, f(t)}{(x^2 - t^2)^{1-\alpha}}\, dt, \qquad (3.216)$$

which will be easily confirmed by a change of variable $t^2 = u$.

Example 3.19. Given a roller coaster of a definite shape, we can calculate the time $f(y)$ such that a body starting from a point $P(x, y)$ reaches the goal $O(0,0)$. Abel proposed the inverse problem (1823) of determining the shape of the coaster rail $\phi(y)$ such that a body reaches the target point O after a specified time.

If the frictional force on the body is negligible, the law of energy conservation holds. Furthermore, if the body is initially at rest at the starting point $P(x, y)$, the speed of the body passing an arbitrary point $Q(\xi, \eta)$ on the rail is $v = \sqrt{2g(y - \eta)}$, where the acceleration of gravity g is assumed to be constant. The time df necessary to pass through the line segment ds along the slope in the neighborhood of point Q is $df = ds/v$, whereas $ds = \sqrt{1 + (d\xi/d\eta)^2}\, d\eta$, so that the time necessary to reach the point O ($y = 0$) starting from the point P is

$$f(y) = \int_y^0 \frac{\sqrt{1 + (d\xi/d\eta)^2}}{\sqrt{2g(y - \eta)}}\, d\eta. \tag{3.217}$$

This is the Abel's equation with $\alpha = 1/2$, in which x and t are replaced by y and η, respectively, and

$$\phi(\eta) = -\frac{\sqrt{1 + (d\xi/d\eta)^2}}{\sqrt{2g}}.$$

The solution is accordingly given by

$$\phi(y)\left(= -\frac{\sqrt{1 + (dx/dy)^2}}{\sqrt{2g}}\right) = \frac{1}{\pi}\frac{d}{dy}\int_0^y \frac{f(\eta)}{\sqrt{y - \eta}}\, d\eta. \tag{3.218}$$

$\triangleleft$

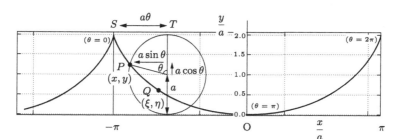

Fig. 3.12   Cycloid ($a = A/2$).

As an example, we consider the special case in which the time to reach the point O does not depend on the starting point P, i.e., $f(y) = T_0$ (constant). Then, Eq. (3.218) becomes

$$\phi(y) = \frac{T_0}{\pi}\frac{d}{dy}\int_0^y \frac{1}{\sqrt{y - \eta}}\, d\eta = \frac{T_0}{\pi\sqrt{y}},$$

or

$$\frac{dy}{dx} = \pm\sqrt{\frac{y}{A - y}}, \qquad A = \frac{2gT_0^2}{\pi^2}, \qquad (3.219)$$

which is the same as Eq. (1.76) of Example 1.19, except for the difference in the choice of the y axis. Considering the physical configuration (Fig. 3.12), we adopt the $-$ sign from the $\pm$ sign before the square root. We solve the Eq. (3.219), by putting $y = A\cos^2(\theta/2)$. In terms of the parameter θ, the solution is given by

$$x = \frac{A}{2}(\theta - \sin\theta) + C, \quad y = \frac{A}{2}(1 + \cos\theta), \qquad (3.220)$$

where C is a constant of integration. Given the initial condition, *e.g.*, $x = x_0$, $y = y_0$ at $t = 0$, then the second equation of (3.220) determines $\theta = \theta_0$, which is substituted for the first equation to determine the constant C. For instance, if the initial position S is $(-\pi a, 2a)$, at which we choose $\theta_0 = 0$, then we have $A = 2a$ and $C = -\pi a$.

The curve described by Eq. (3.220) is a cycloid, which is a periodic function of period $\theta = 2\pi$. If a body is released from an arbitrary point on the frictionless concave surface of this form, the body makes an oscillatory motion about the lowest position O ($x = y = 0$ in Fig. 3.12). The period of the latter is kept constant regardless of the initial position of the body, and hence, is independent of the amplitude of the oscillation, which is well-known as a cycloidal pendulum.

A number of problems and solutions related to the present chapter can be found in *e.g.* [DuChateau and Zachmann (1986)].

Appendix A

A.1 Gamma Function and Beta Function

A.1.1 *Gamma function*

The gamma function is defined by

$$\Gamma(z) = \int_0^\infty e^{-t} t^{z-1}\, dt \qquad \text{for} \quad \text{Re}\, z > 0. \tag{A.1}$$

Integrating by parts:

$$\int_0^\infty e^{-t} t^z dt = z \int_0^\infty e^{-t} t^{z-1}\, dt,$$

we have

$$\Gamma(z+1) = z\Gamma(z). \tag{A.2}$$

If z is an integer n, it gives

$$\Gamma(n+1) = n!, \qquad (n = 1, 2, 3, \cdots), \tag{A.3}$$

where $0! = 1$ is assumed.

We leave the details on the gamma function to [Morse and Feshbach (1953)]($\S$4.5), [Abramowitz and Stegun (1964)]($\S$6.1), [Carrier, Krook and Pearson (1966)]($\S$5.1), [Arfken (1985)]($\S$10.1), *etc.*, and mention a few fundamental properties that are used in this textbook, among which are

$$\Gamma(z)\Gamma(1-z) = (z-1)!(-z)! = \frac{\pi}{\sin \pi z}, \tag{A.4}$$

$$\Gamma\left(\frac{1}{2}+z\right)\Gamma\left(\frac{1}{2}-z\right) = \frac{\pi}{\cos \pi z}, \tag{A.5}$$

$$\Gamma(2z) = \frac{2^{2z-1}}{\sqrt{\pi}} \Gamma(z)\Gamma\left(z+\frac{1}{2}\right). \tag{A.6}$$

From these relations, we can easily reproduce the result $\Gamma(1/2) = \sqrt{\pi}$, which can, of course, be obtained by a direct calculation.

A.1.2 Beta function

The beta function $B(p,q)$ is defined by

$$B(p,q) = \int_0^1 t^{p-1}(1-t)^{q-1}\mathrm{d}t, \qquad (\mathrm{Re}\,p > 0, \mathrm{Re}\,q > 0). \tag{A.7}$$

By a transformation of variables $t = \tau/(1+\tau)$ or $t = \sin^2\theta$, it is also expressed as

$$B(p,q) = \int_0^\infty \frac{\tau^{p-1}}{(1+\tau)^{p+q}}\,\mathrm{d}\tau = 2\int_0^{\pi/2} \cos^{2p-1}\theta \sin^{2q-1}\theta\,\mathrm{d}\theta.$$

The latter has a relation to the gamma function:

$$B(p,q) = \frac{\Gamma(p)\Gamma(q)}{\Gamma(p+q)} = B(q,p). \tag{A.8}$$

For further details, see references *e.g.*, [Carrier, Krook and Pearson (1966)]($5.1), [Arfken (1985)]($10.4), *etc.*

A.2 Bessel Functions

A.2.1 Bessel functions and Neumann functions

In a circular cylindrical coordinate system (ρ, φ, z), the Laplace equation $\triangle\Phi = 0$ is given by

$$\frac{1}{\rho}\frac{\partial}{\partial\rho}\left(\rho\frac{\partial\Phi}{\partial\rho}\right) + \frac{1}{\rho^2}\frac{\partial^2\Phi}{\partial\varphi^2} + \frac{\partial^2\Phi}{\partial z^2} = 0. \tag{A.9}$$

If we assume a solution in the form (separation-of-variables type) $\Phi(\rho,\varphi,z) = f(\rho)\exp(in\varphi \pm kz)$, the equation for f becomes

$$\frac{\mathrm{d}^2 f}{\mathrm{d}\rho^2} + \frac{1}{\rho}\frac{\mathrm{d}f}{\mathrm{d}\rho} + \left(k^2 - \frac{n^2}{\rho^2}\right)f = 0, \tag{A.10}$$

or by a further transformation of the variable $k\rho = x$,

$$\frac{\mathrm{d}^2 f}{\mathrm{d}x^2} + \frac{1}{x}\frac{\mathrm{d}f}{\mathrm{d}x} + \left(1 - \frac{n^2}{x^2}\right)f = 0. \tag{A.11}$$

The solution of Eq. (A.11), denoted by $J_n(x)$, is called the n-th order Bessel function of the first kind.

In the following, we show some properties of the Bessel functions, (for details, refer to *e.g.*, [Morse and Feshbach (1953)]($5.3), [Abramowitz and Stegun (1964)](Chaps. 9 and 10), [Carrier, Krook and Pearson (1966)]($5.5), [Arfken (1985)](Chap. 11), *etc.*).

a. Power series expansion at origin $x = 0$

The power series solution is given by the well-known procedures used in an ODE. Expansion at the origin $x = 0$ is

$$J_0(x) = 1 - \frac{x^2}{4} + \frac{x^4}{64} - \cdots , \qquad J_1(x) = \frac{x}{2} - \frac{x^3}{16} + \frac{x^5}{384} - \cdots , \qquad \text{(A.12)}$$

and in general

$$J_n(x) = \sum_{m=0}^{\infty} \frac{(-1)^m}{(n+m)!m!} \left(\frac{x}{2}\right)^{2m+n} . \qquad \text{(A.13)}$$

The latter expansion holds for a real number ν in place of an integer n.

Note that if ν is not an integer, $J_{-\nu}(x)$ is a solution independent of $J_\nu(x)$. In contrast, if ν is an integer, $J_{-n}(x)$ is not independent of $J_n(x)$. To see this, we formally put $n \to -n$ in Eq. (A.13):

$$J_{-n}(x) = \sum_{m=0}^{\infty} \frac{(-1)^m}{(m-n)!m!} \left(\frac{x}{2}\right)^{2m-n} . \qquad \text{(A.14)}$$

In the latter, however, factorials of the negative integers appear in the denominator, such as $(-n)!, (1-n)!, \cdots , (-1)!$ for $m = 0, 1, 2, \cdots , n-1$. We know, however, that the factorial of the negative integers becomes plus or minus infinite because negative integers are multiplied indefinitely:

$$(-1)! = (-1)(-2)(-3) \cdots , \qquad (-2)! = (-2)(-3)(-4) \cdots , \qquad \cdots .$$

As a result, the coefficients of the series expansion (A.14) with $m = 0 \sim n-1$ become zero, so that we have

$$J_{-n}(x) = \sum_{m=n}^{\infty} \frac{(-1)^m}{(m-n)!m!} \left(\frac{x}{2}\right)^{2m-n}$$

$$= \sum_{k=0}^{\infty} \frac{(-1)^{n+k}}{k!(n+k)!} \left(\frac{x}{2}\right)^{n+2k} = (-1)^n J_n(x).$$

(From the second to the third terms, we put $m = n + k$, and compare it with Eq. (A.13).) This equation reveals that $J_{-n}(x)$ and $J_n(x)$ are not independent of each other for an integer n.

We, therefore, define the second solution independent of $J_n(x)$:

$$N_\nu(x) = \frac{J_\nu(x) \cos \nu\pi - J_{-\nu}(x)}{\sin \nu\pi}, \qquad N_n(x) = \lim_{\nu \to n} N_\nu(x), \qquad \text{(A.15)}$$

which is called the n-th order Bessel function of the second kind, or the **Neumann function**.

The power series expansions of the Neumann functions are

$$N_0(x) = \frac{2}{\pi} (\log x - \log 2 + \gamma) J_0(x) + \frac{2}{\pi} \left(\frac{x^2}{4} - \frac{3x^4}{128} + \cdots \right), \tag{A.16}$$

$$N_1(x) = \frac{2}{\pi} (\log x - \log 2 + \gamma) J_1(x) - \frac{2}{\pi} \left(\frac{1}{x} + \frac{x}{4} - \frac{5x^5}{64} + \cdots \right), \tag{A.17}$$

and in general

$$N_n(x) = \frac{2}{\pi} (\log x - \log 2 + \gamma) J_n(x)$$

$$- \frac{1}{\pi} \sum_{m=0}^{\infty} \frac{(-1)^m}{(n+m)!m!} \left[\sum_{l=1}^{m} \frac{1}{l} + \sum_{l=1}^{m+l} \frac{1}{l} \right] \left(\frac{x}{2} \right)^{2m+n}$$

$$- \frac{1}{\pi} \sum_{m=0}^{n-1} \frac{(n-m-1)!}{m!} \left(\frac{x}{2} \right)^{2m-n}, \tag{A.18}$$

where $\gamma = 0.5772156649\cdots$ is the Euler constant. We show these functions in Fig. A.1(a).

b. Asymptotic behavior

For a half integer n, J_ν is given in terms of the elementary functions, such as

$$J_{1/2}(x) = \sqrt{\frac{2}{\pi x}} \sin x, \quad J_{-1/2}(x) = \sqrt{\frac{2}{\pi x}} \cos x,$$

$$J_{3/2}(x) = \sqrt{\frac{2}{\pi x}} \left(-\cos x + \frac{\sin x}{x} \right), \quad J_{-3/2}(x) = -\sqrt{\frac{2}{\pi x}} \left(\sin x + \frac{\cos x}{x} \right), \cdots. \tag{A.19}$$

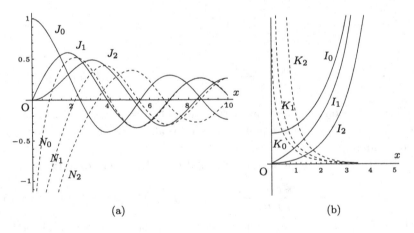

(a)   (b)

Fig. A.1   Bessel functions (a) J_n, N_n, and (b) I_n, K_n.

These functions show an oscillatory behavior with their extremum values decreasing as $x^{-1/2}$ with the increase of x. The zero points of the pair of J_ν and $J_{-\nu}$ appear alternately. The asymptotic behaviors of the Bessel functions of order n are given by

$$J_n(x) \sim \sqrt{\frac{2}{\pi x}} \cos\left(x - \frac{(2n+1)\pi}{4}\right), \tag{A.20}$$

$$N_n(x) \sim \sqrt{\frac{2}{\pi x}} \sin\left(x - \frac{(2n+1)\pi}{4}\right) \sim J_{n+1}(x), \tag{A.21}$$

for $x \gg 1$.

c. Orthogonality and zero points

If we denote the i-th zero of $J_n(x)$ by $\lambda_i^{(n)}$ $(0 < \lambda_1^{(n)} < \lambda_2^{(n)} < \cdots)$, the relation

$$\int_0^1 x J_n(\lambda_i^{(n)} x) J_n(\lambda_j^{(n)} x)\, dx = \frac{1}{2}[J_{n+1}(\lambda_i^{(n)})]^2\, \delta_{ij} \tag{A.22}$$

holds. By the use of this relation, the following expansion in the interval $[0, a]$ is developed:

$$F(x) = \sum_{i=1}^{\infty} c_{in} J_n\left(\lambda_i^{(n)} \frac{x}{a}\right),$$

where

$$c_{in} = \frac{2}{a^2[J_{n+1}(\lambda_i^{(n)})]^2} \int_0^a x F(x) J_n\left(\lambda_i^{(n)} \frac{x}{a}\right) dx, \tag{A.23}$$

which is called the **Fourier–Bessel expansion**. This expansion is useful for the boundary value problems specified on the circular cylinder.

d. Generating function

The Laurent expansions of the function $\exp\left[\frac{x}{2}\left(\zeta - \frac{1}{\zeta}\right)\right]$ around the essential singularity point $\zeta = 0$ gives

$$\exp\left[\frac{x}{2}\left(\zeta - \frac{1}{\zeta}\right)\right] = \sum_{n=-\infty}^{\infty} \zeta^n J_n(x). \tag{A.24}$$

The l.h.s. of Eq. (A.24) is called the generating function of the Bessel functions, which is easily obtained by expanding the l.h.s. as follows,

$$\exp\left[\frac{x}{2}\left(\zeta - \frac{1}{\zeta}\right)\right] = \exp\left(\frac{x\zeta}{2}\right) \exp\left(-\frac{x}{2\zeta}\right)$$

$$= \sum_{m=0}^{\infty} \frac{1}{m!} \left(\frac{x\zeta}{2}\right)^m \sum_{n=0}^{\infty} \frac{1}{n!} \left(-\frac{x}{2\zeta}\right)^n = \sum_{m=0}^{\infty} \sum_{n=0}^{\infty} \frac{1}{m! \, n!} \frac{(-1)^n}{2^{m+n}} x^{m+n} \zeta^{m-n}$$

$$= \sum_{k=-\infty}^{\infty} \zeta^k \sum_{n=0}^{\infty} \frac{(-1)^n}{n! \, (k+n)!} \left(\frac{x}{2}\right)^{k+2n},$$

and is compared with Eq. (A.13).

e. Integral representation

In the Laurent series expansion using the generating function (A.24), the Bessel function $J_n(x)$ is the coefficient of ζ^n, so that it is obtained by the integral (see Cauchy's integral formula, or Goursat's theorem)

$$J_n(x) = \frac{1}{2\pi i} \int_C \frac{1}{\zeta^{n+1}} \exp\left[\frac{x}{2}\left(\zeta - \frac{1}{\zeta}\right)\right] d\zeta,$$

where C is an arbitrary closed contour that encircles the singular point $\zeta = 0$. Further transformation using $\zeta = \exp(i\theta)$ yields

$$J_n(x) = \frac{1}{2\pi} \int_\alpha^{2\pi+\alpha} e^{i(x \sin\theta - n\theta)} d\theta, \qquad (A.25)$$

where α is an arbitrary real constant.

f. Recurrence formulae

The Bessel functions have the following recurrence formulae:

$$J_{n-1}(x) = \frac{2n}{x} J_n(x) - J_{n+1}(x), \quad J_{n+1}(x) = \frac{2n}{x} J_n(x) - J_{n-1}(x). \quad (A.26)$$

By the use of the above relations, higher order Bessel functions are given by J_0 and J_1. For example,

$$J_2(x) = \frac{2}{x} J_1(x) - J_0(x),$$

$$J_3(x) = \frac{4}{x} J_2(x) - J_1(x) = \left(\frac{8}{x^2} - 1\right) J_1(x) - \frac{4}{x} J_0(x), \quad \cdots . \quad (A.27)$$

The Bessel functions also have the recurrence formulae on the differentiation:

$$\left(\frac{d}{dx} - \frac{n}{x}\right) J_n(x) = -J_{n+1}(x), \quad \left(\frac{d}{dx} + \frac{n}{x}\right) J_n(x) = J_{n-1}(x), \quad (A.28)$$

which increase or decrease the order of J_n by 1, so that the above operators are called the ladder operators (more precisely, the former and the latter of Eq. (A.28) are called a raising operator and a lowering operator, respectively). By the use of these relations, we have, *e.g.*,

$$J_0' = -J_1, \quad J_1' = J_0 - \frac{1}{x}J_1, \quad \cdots . \tag{A.29}$$

Furthermore, by combining the two types of recurrence formulae (A.28) and (A.27), we have

$$J_2' = J_1 - \frac{2}{x}J_2 = \frac{2}{x}J_0 + \left(1 - \frac{4}{x^2}\right)J_1, \quad \cdots , \tag{A.30}$$

where we denote $df(x)/dx = f'$ for brevity.

A.2.2 *Hankel functions*

Linear combinations of J_n and N_n mentioned above

$$H_n^{(1)}(x) = J_n(x) + iN_n(x), \qquad H_n^{(2)}(x) = J_n(x) - iN_n(x) \tag{A.31}$$

are called the **Hankel functions** or the Bessel functions of the third kind. These functions have a resemblance to the Euler's formula $\exp(\pm ix) = \cos x \pm i \sin x$, and their asymptotic behaviors at $x \gg 1$ are given by

$$\left.\begin{matrix} H_n^{(1)} \\ H_n^{(2)} \end{matrix}\right\} \sim \sqrt{\frac{2}{\pi x}} \exp\left[\pm i\left(x - \frac{(2n+1)\pi}{4}\right)\right]. \tag{A.32}$$

Recurrence formulae are the same as those for J_n and N_n. For a pure imaginary argument, the relation

$$H_n^{(1)}(ix) = \frac{2}{\pi i^{n+1}} K_n(x) \tag{A.33}$$

holds, where K_n is the n-th order "modified Bessel function" of the second kind to be shown in §§A.2.4.

A.2.3 *Spherical Bessel functions*

The Bessel functions with a half-integral order are called the spherical Bessel functions. The ODE that governs the latter is

$$\frac{d^2 f}{dx^2} + \frac{2}{x}\frac{df}{dx} + \left(1 - \frac{n(n+1)}{x^2}\right)f = 0. \tag{A.34}$$

The solutions of this ODE are j_n, n_n, $h_n^{(1)}$, $h_n^{(2)}$, which are respectively defined by J_n, N_n, $H_n^{(1)}$, $H_n^{(2)}$ with their order n replaced by $n + 1/2$. For example,

$$j_n(x) = \sqrt{\frac{\pi}{2x}}J_{n+1/2}(x), \qquad n_n(x) = \sqrt{\frac{\pi}{2x}}N_{n+1/2}(x), \tag{A.35}$$

$$h_n^{(1)}(x) = \sqrt{\frac{\pi}{2x}}H_{n+1/2}^{(1)}(x), \qquad h_n^{(2)}(x) = \sqrt{\frac{\pi}{2x}}H_{n+1/2}^{(2)}(x). \tag{A.36}$$

Combined with Eq. (A.19), we have

$$j_0(x) = \frac{\sin x}{x}, \qquad j_1(x) = -\frac{\cos x}{x} + \frac{\sin x}{x^2},$$

$$j_2(x) = \left(-\frac{1}{x} + \frac{3}{x^3}\right)\sin x - \frac{3}{x^2}\cos x, \ \ldots , \qquad (A.37)$$

$$n_0(x) = -\frac{\cos x}{x}, \qquad n_1(x) = -\frac{\sin x}{x} - \frac{\cos x}{x^2},$$

$$n_2(x) = \left(\frac{1}{x} - \frac{3}{x^3}\right)\cos x - \frac{3}{x^2}\sin x, \ \ldots . \qquad (A.38)$$

With the use of Eq. (A.31), we also have

$$h_0^{(1)}(x) = -\frac{i}{x}\,e^{ix}, \qquad h_1^{(1)}(x) = -\left(\frac{1}{x} + \frac{i}{x^2}\right)e^{ix}, \ \ldots ,$$

$$h_0^{(2)}(x) = \frac{i}{x}\,e^{-ix}, \qquad h_1^{(2)}(x) = \left(-\frac{1}{x} + \frac{i}{x^2}\right)e^{-ix}, \ \ldots . \qquad (A.39)$$

These functions show an oscillatory behavior with their extremum values decreasing as $1/x$ with the increase of x. The zero points alternate between the pair of relevant functions.

A.2.4 *Modified Bessel functions*

If we assume a solution of Eq. (A.9) in the form $\Phi(\rho, \varphi, z) = f(\rho)\exp(in\varphi \pm ikz)$, and put $k\rho = x$, then the equation satisfied by f becomes

$$\frac{d^2 f}{dx^2} + \frac{1}{x}\frac{df}{dx} - \left(1 + \frac{n^2}{x^2}\right)f = 0. \qquad (A.40)$$

The solutions, denoted by $I_n(x)$ and $K_n(x)$, are called the n-th order modified Bessel functions of the first kind and the second kind, respectively. Formally, they are the same as the previously mentioned Bessel functions (§§A.2.1), in which k is replaced by ik. We show some of these functions in Fig. A.1(b).

a. Behavior near $x = 0$

$$I_0(x) = 1 + \frac{x^2}{4} + \frac{x^4}{64} + \cdots , \qquad I_1(x) = \frac{x}{2} + \frac{x^3}{16} + \frac{x^5}{384} + \cdots ,$$

$$K_0(x) = -(\log x - \log 2 + \gamma)\,I_0(x) + \frac{x^2}{4} + \frac{3x^4}{128} + \cdots ,$$

$$K_1(x) = \frac{1}{x} + (\log x - \log 2 + \gamma)\,I_1(x) - \frac{x}{4} - \frac{5x^3}{64} + \cdots , \qquad (A.41)$$

$$I_n(x) \approx \frac{1}{\Gamma(n+1)}\left(\frac{x}{2}\right)^n, \qquad K_n(x) \approx \frac{\Gamma(n)}{2}\left(\frac{2}{x}\right)^n .$$

b. Half-integer order n

In this case, they are given by elementary functions. For example,

$$I_{1/2}(x) = \sqrt{\frac{2}{\pi x}} \sinh x, \quad I_{-1/2}(x) = \sqrt{\frac{2}{\pi x}} \cosh x,$$

$$I_{3/2}(x) = \sqrt{\frac{2}{\pi x}} \left(\cosh x - \frac{\sinh x}{x} \right),$$

$$I_{-3/2}(x) = \sqrt{\frac{2}{\pi x}} \left(\sinh x - \frac{\cosh x}{x} \right), \quad \cdots , \tag{A.42}$$

$$K_{1/2}(x) = K_{-1/2}(x) = \sqrt{\frac{\pi}{2x}} e^{-x},$$

$$K_{3/2}(x) = K_{-3/2}(x) = \sqrt{\frac{\pi}{2x}} \left(1 + \frac{1}{x} \right) e^{-x}, \quad \cdots .$$

c. Asymptotic expansion at a large x

$$I_n(x) \sim \frac{e^x}{\sqrt{2\pi x}} \left(1 - \frac{4n^2 - 1}{8x} + \cdots \right),$$

$$K_n(x) \sim \frac{\sqrt{\pi} e^{-x}}{\sqrt{2x}} \left(1 + \frac{4n^2 - 1}{8x} + \cdots \right). \tag{A.43}$$

A.3 Legendre Functions

In the spherical coordinate system (r, θ, φ), the Laplace equation $\triangle \Phi = 0$ is given by

$$\frac{1}{r^2} \frac{\partial}{\partial r} \left(r^2 \frac{\partial \Phi}{\partial r} \right) + \frac{1}{r^2 \sin \theta} \frac{\partial}{\partial \theta} \left(\sin \theta \frac{\partial \Phi}{\partial \theta} \right) + \frac{1}{r^2 \sin^2 \theta} \frac{\partial^2 \Phi}{\partial \varphi^2} = 0. \tag{A.44}$$

If we assume a solution in the form $\Phi(r, \theta, \varphi) = r^n Y_n(\theta, \varphi)$ $(n = 0, 1, 2, \cdots)$, then Y_n satisfies the following equation:

$$\frac{1}{\sin \theta} \frac{\partial}{\partial \theta} \left(\sin \theta \frac{\partial Y_n}{\partial \theta} \right) + \frac{1}{\sin^2 \theta} \frac{\partial^2 Y_n}{\partial \varphi^2} + n(n+1) Y_n = 0. \tag{A.45}$$

The solution $Y_n(\theta, \varphi)$ is called the n-th order **spherical surface harmonic function**. Furthermore, if the variables in Y_n are separated as $Y_n(\theta, \varphi) = \Theta(\theta) \exp(\pm im\varphi)$, then Θ satisfies the ODE:

$$\frac{1}{\sin \theta} \frac{d}{d\theta} \left(\sin \theta \frac{d\Theta}{d\theta} \right) + \left[n(n+1) - \frac{m^2}{\sin^2 \theta} \right] \Theta = 0. \tag{A.46}$$

The solutions of this equation, $P_n^m(\cos \theta)$ and $Q_n^m(\cos \theta)$, are called the **associated Legendre functions** of the first kind and the second kind, respectively. Here, $|m| = 1, 2, 3, \cdots, n$, and $P_n^m, Q_n^m = 0$ for $|m| > n$. For a given n, we have linearly independent $2n + 1$ functions with $m = 0, \pm 1, \pm 2, \cdots, \pm n$ [or $\cos m\varphi$, $\sin m\varphi$ corresponding to $\exp(\pm im\varphi)$].

In the following, we show some basic properties of the Legendre and the associated Legendre functions (for further details, refer to e.g., [Morse and Feshbach (1953)](§5.2), [Abramowitz and Stegun (1964)](Chap. 8), [Carrier, Krook and Pearson (1966)](§5.4), [Arfken (1985)](Chap. 12), etc.).

A.3.1 *Legendre functions*

Let us first consider the case $m = 0$. The solution is axisymmetric, and the ODE for Θ is

$$\frac{1}{\sin\theta}\frac{d}{d\theta}\left(\sin\theta\frac{d\Theta}{d\theta}\right) + n(n+1)\Theta = 0, \tag{A.47}$$

or

$$(1 - x^2)\frac{d^2\Theta}{dx^2} - 2x\frac{d\Theta}{dx} + n(n+1)\,\Theta = 0, \tag{A.48}$$

where $\cos\theta = x$. The solutions $P_n(\cos\theta)$ and $Q_n(\cos\theta)$ are called the **Legendre functions** of the first kind and the second kind, respectively. As shown later, $P_n(\cos\theta)$ is finite but $Q_n(\cos\theta)$ becomes infinite at $\theta = \pm\pi$, so that the solution of $\Delta\Phi = 0$ that is axisymmetric and regular $r = 0$ is given by

$$\Phi = \sum_{n=0}^{\infty} A_n r^n P_n(\cos\theta). \tag{A.49}$$

a. Power series expansion at $x = 0$

The Legendre function of the first kind $P_n(x)$ with an integer n is given by the polynomials, as shown in §§2.3.3b:

$$P_0(x) = 1, \; P_1(x) = x, \; P_2(x) = \frac{1}{2}(3x^2 - 1), \; P_3(x) = \frac{1}{2}(5x^3 - 3x), \cdots. \tag{A.50}$$

In general, they are given by the **Rodrigues' formula**:

$$P_n(x) = \frac{1}{2^n n!}\frac{d^n}{dx^n}(x^2 - 1)^n, \tag{A.51}$$

or

$$P_n(x) = \frac{1}{2^n}\sum_{k=0}^{[n/2]}(-1)^k\frac{(2n - 2k)!}{k!(n - k)!(n - 2k)!}x^{n-2k}. \tag{A.52}$$

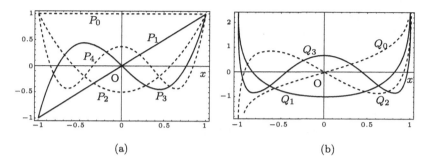

Fig. A.2   (a) Legendre polynomials $P_n(x)$ ($n = 0, 1, 2, 3, 4$), and (b) Legendre functions of the second kind $Q_n(x)$ ($n = 0, 1, 2, 3, 4$).

Having these properties, $P_n(x)$ is called the **Legendre polynomials**. As is evident, the relation $P_n(-x) = (-1)^n P_n(x)$ holds. We first show a few of $P_n(x)$ in Fig. A.2(a).

In contrast, the Legendre functions of the second kind are

$$Q_0(x) = \frac{1}{2}\log\frac{1+x}{1-x}, \qquad Q_1(x) = \frac{x}{2}\log\frac{1+x}{1-x} - 1, \qquad (A.53)$$

$$Q_2(x) = \frac{1}{4}(3x^2 - 1)\log\frac{1+x}{1-x} - \frac{3}{2}x, \quad \cdots, \qquad (A.54)$$

or in general

$$Q_n(x) = \frac{1}{2}P_n(x)\log\frac{1+x}{1-x} - W_{n-1}(x), \qquad (A.55)$$

where

$$W_{-1} = 0, \qquad W_{n-1}(x) = \sum_{s=0}^{[(n-1)/2]}\frac{(2n-4s-1)}{(2s+1)(n-s)}P_{n-2s-1}(x). \qquad (A.56)$$

They have logarithmic singularities at $x = \pm 1$. We show a few of $Q_n(x)$ in Fig. A.2(b).

b. Orthogonality

The Legendre polynomials $P_n(x)$ are orthogonal to each other in the interval $[-1, 1]$, and are given by

$$\int_{-1}^{1} P_m(x)P_n(x)\,\mathrm{d}x = \frac{2}{2n+1}\delta_{mn}, \qquad (A.57)$$

so that $\sqrt{(2n+1)/2}\,P_n(x)$ ($n = 0, 1, 2, \cdots$) constitutes an orthonormal set.

c. Generating function

The Legendre polynomials $P_n(x)$ are given as coefficients of the power series expansion of

$$G(x,t) \equiv \frac{1}{\sqrt{1 - 2xt + t^2}} = \sum_{n=0}^{\infty} P_n(x)\, t^n, \qquad (A.58)$$

so that the function $G(x,t)$ is the generating function of $P_n(x)$. Using the latter, the reciprocal of the distance R between two points A and B is expanded in the series

$$\frac{1}{R} = \begin{cases} \displaystyle\sum_{n=0}^{\infty} \frac{r^n}{r'^{n+1}} P_n(\cos\theta) & (r' > r) \\[4mm] \displaystyle\sum_{n=0}^{\infty} \frac{r'^n}{r^{n+1}} P_n(\cos\theta) & (r' < r) \end{cases} . \qquad (A.59)$$

Here, r and r' are the distances to the points A and B, respectively, from the origin O, θ is the angle between the line OA and OB, and $R = \sqrt{r^2 - 2rr'\cos\theta + r'^2}$.

A.3.2   *Associated Legendre functions*

The solutions of the associated Legendre differential equation, $P_n^m(x)$ and $Q_n^m(x)$, are given by

$$P_n^m(x) = (1 - x^2)^{m/2} \frac{\mathrm{d}^m}{\mathrm{d}x^m} P_n(x), \quad Q_n^m(x) = (1 - x^2)^{m/2} \frac{\mathrm{d}^m}{\mathrm{d}x^m} Q_n(x), \qquad (A.60)$$

which are called the **associated Legendre functions** of the first kind and the second kind, respectively. The first few of them are

$$P_0^0(x) = P_0(x), \qquad P_1^0(x) = P_1(x), \qquad P_1^1(x) = \sqrt{1 - x^2} = \sin\theta,$$

$$P_2^0(x) = P_2(x), \qquad P_2^1(x) = 3x\sqrt{1 - x^2} = 3\sin\theta\cos\theta = \frac{3}{2}\sin 2\theta,$$

$$P_2^2(x) = 3(1 - x^2) = 3\sin^2\theta = \frac{3}{2}(1 - \cos 2\theta), \quad \cdots ,$$

$$(A.61)$$

where $x = \cos\theta$. The general expression for $P_n^{-m}(x)$ is given by

$$P_n^{-m}(x) = (-1)^m \frac{(n - m)!}{(n + m)!} P_n^m(x), \qquad (m > 0), \qquad (A.62)$$

and the orthogonality relation is

$$\int_{-1}^{1} P_l^m(x) P_n^m(x)\,\mathrm{d}x = \frac{2}{2n + 1} \frac{(n + m)!}{(n - m)!} \delta_{ln}. \qquad (A.63)$$

A.3.3 *Spherical surface harmonic functions*

We have already referred to the solution $Y_n(\theta, \varphi)$ of $\triangle \Phi = 0$ in terms of the spherical coordinate system, in which the dependence on r is separated. In other words, this implies that $Y_n(\theta, \varphi)$ gives a harmonic function on the spherical surface $r = $ constant, so that it is called the **spherical surface harmonic function**. We show the orthonormal set of functions using the solutions that are finite on the spherical surface $P_n^m(\cos\theta)$ and the function $e^{\pm im\varphi}$. A customary description of the functions $Y_n(\theta, \varphi)$ used to clarify the dependence on θ, φ is $Y_{n,m}(\theta, \varphi)$:

$$Y_{n,m}(\theta, \varphi) = \tilde{P}_n^m(\cos\theta)\,\tilde{\Phi}_m(\varphi), \qquad (m = 0, \pm 1, \pm 2, \cdots, \pm n), \qquad (A.64)$$

where

$$\tilde{P}_n^m(\cos\theta) = \sqrt{\frac{2n+1}{2}\frac{(n-m)!}{(n+m)!}}\,P_n^m(\cos\theta), \quad \tilde{\Phi}_m(\varphi) = \frac{1}{\sqrt{2\pi}}\,e^{im\varphi}. \quad (A.65)$$

The first few of them are

$$Y_{0,0}(\theta, \varphi) = \frac{1}{\sqrt{4\pi}},$$

$$Y_{1,0}(\theta, \varphi) = \sqrt{\frac{3}{4\pi}}\cos\theta, \qquad Y_{1,\pm 1}(\theta, \varphi) = \pm\sqrt{\frac{3}{8\pi}}\sin\theta\,e^{\pm i\varphi},$$

$$Y_{2,0}(\theta, \varphi) = \frac{1}{2}\sqrt{\frac{5}{4\pi}}(3\cos^2\theta - 1), \quad Y_{2,\pm 1}(\theta, \varphi) = \pm\sqrt{\frac{15}{8\pi}}\sin\theta\cos\theta\,e^{\pm i\varphi},$$

$$Y_{2,\pm 2}(\theta, \varphi) = \frac{1}{2}\sqrt{\frac{15}{8\pi}}\sin^2\theta\,e^{\pm 2i\varphi},$$

$$Y_{3,0}(\theta, \varphi) = \frac{1}{2}\sqrt{\frac{7}{4\pi}}\cos\theta\,(5\cos^2\theta - 3), \quad \cdots.$$

$$(A.66)$$

Bibliography

Abramowitz, M. and Stegun, I. A. (eds.) (1964). *Handbook of Mathematical Functions with Formulas, Graphs, and Mathematical Tables*, (U.S. Department of Commerce, National Bureau of Standards, Applied Mathematics Series - 55), (also Dover).

Arfken, G. (1985). *Mathematical Methods for Physicists*, (Academic Press).

Beiser, A. (1995). *Concepts of Modern Physics*, 5th edn., (McGraw-Hill).

Brown, J. W. and Churchill, R. V. (2009). *Complex Variables and Applications*, 8th edn., (McGraw-Hill).

Byron, F. W. and Fuller, R. W. (1969). *Mathematics of Classical and Quantum Physics*, (Addison-Wesley).

Carrier, G. F., Krook, M. and Pearson, C. E. (1966). *Functions of a Complex Variable*, (McGraw-Hill).

DuChateau, P. and Zachmann, D. W. (1986). *Theory and Problems of Partial Differential Equations*, (McGraw-Hill).

Erdélyi, A. (ed.), Magnus, W., Oberhettinger, F. and Tricomi, F. G. (1954). *Tables of Integral Transforms*, Vol. I, II (McGraw-Hill).

Fujiwara, T. (2013). *Complex Analysis*, (Mathematics/UTokyo Engineering Course), (University of Tokyo).

John, F. (1982). *Partial Differential Equations (Applied Mathematical Sciences)*, 4th edn., (Springer).

MacLachlan, N. W. (1962). *Laplace transforms, and their applications to differential equations*, (Dover).

Morse, P. M. and Feshbach, H. (1953). *Methods of Theoretical Physics*, (McGraw-Hill).

Schiff, L. I. (1949). *Quantum Mechanics*, (McGraw-Hill).

Sneddon, I. N. (1966). *Mixed Boundary Value Problems in Potential Theory*, (North-Holland).

Titchmarsh, E. C. (1948). *Introduction to the Theory of Fourier Integrals*, 2nd edn., (Oxford University Press).

Index

Abel's equation, 198
adjoint differential operator, 95, 123
adjoint Green's function, 106, 125
associated Laguerre polynomials, 84
associated Legendre functions, 84,
    209, 212
auxiliary equations, 18, 20, 32, 47

Bessel functions, 84, 91, 188, 202
Bessel inequality, 85
beta function, 181, 202
bipolar cylinder coordinates, 79
bispherical coordinates, 79
boundary condition, 4, 52
boundary-value problem, 115, 116,
    125, 127, 130, 132
Brachistochrone, 39
Bromwich integral, 151, 174

canonical equation, 43
canonical transform, 43
Cartesian (rectangular) coordinates,
    66, 67
Cauchy condition, 60, 81
Cauchy's problem, 21, 60
causality, 65, 107
characteristic curve, 9, 10, 14, 17, 61
characteristic differential equation,
    8–10, 13, 15, 20
characteristic direction, 14, 62
characteristic polynomial, 172
characteristic strip, 17

Chebyshev polynomials, 84
Clairaut's equation, 25, 36
complementary error function, 126,
    176
complete integral, 23
complete solution, 23, 27
complete system, 83
cone, 5, 6, 20
convergence in the mean, 83
convolution, 159, 160, 171, 173, 180,
    196–198
curl, 67
curve of steepest descent, 39
curvilinear coordinates, 66
cut, 174
cycloid, 41, 200
cylindrical coordinates, 76

d'Alembert's solution, 64
d'Alembert–Stokes' solution, 118
degenerate, 147, 149
delta function, 93, 94, 98, 156, 157
diffusion constant, 53
diffusion equation, 53, 59, 65, 114
diffusion(1D), 53, 59, 110, 123–126,
    166, 174
diffusion(3D), 112
diffusion(circular), 136
Dirichlet condition, 81
Dirichlet's problem, 116, 127, 130,
    132, 163

divergence (div), 67
dual integral equations, 193

eigenfunction, 81, 84
eigenfunction expansion, 83, 85
eigenvalue, 81, 84
elliptic cylinder coordinates, 77
elliptic partial differential equation
  (elliptic PDE), 58
envelope, 15, 27
error function, 126, 176
Euler–Lagrange equation, 38
exact, 29
extremum, 37

Faltung, 159
finite integral transform, 146
Fourier cosine transform, 150, 153
Fourier integral formula, 149
Fourier kernel, 184
Fourier series, 83
Fourier sine transform, 150, 155
Fourier transform, 149, 155
Fourier–Bessel expansion, 93, 136,
  205
Fourier–Bessel integral formula, 152,
  188
Fredholm equation, 194–196
functional, 37
functional relation, 3, 5, 16, 24

gamma function, 153, 201
Gaussian, 126, 153, 155
general integral, 5, 24
general solution, 5, 20, 24, 63, 88,
  118, 134, 137, 147, 165
generalized Green's formula, 96
generating function, 44, 205, 212
gradient (grad), 67
Gram–Schmidt orthogonalization
  method, 89
Green's formula, 96
Green's function, 98, 114–116, 125,
  130, 132, 137

Hamilton's equation, 43

Hamilton's principal function, 46
Hamilton–Jacobi equation, 46
Hamiltonian, 42, 47
Hankel functions, 114, 138, 207
Hankel transform, 152, 187
harmonic function, 54, 117, 118
heat conduction equation, 59, 65
Heaviside function, 95
Helmholtz equation, 65, 114
Helmholtz(axisymmetric), 138
Helmholtz(spherical), 133, 136
Hermite polynomials, 84, 90
Huygens' principle, 122
hyperbolic partial differential
  equation (hyperbolic PDE), 57

indeterminacy, 39, 44, 99
infinite integral transform, 148
initial condition, 10, 53, 63
initial-value problem, 21, 22, 59–61,
  118, 120, 123, 125, 172
integrable condition, 30
integral equation, 194
integral functional, 37, 42
integral representation, 157, 206
integral surface, 8, 59
integral transform, 141
integrating factor, 29

Jacobian, 4, 7, 23, 24, 56, 61
Jordan's lemma, 101, 163, 175

kernel, 141, 143, 148, 182–184

Lagrange's equation of motion, 42
Lagrange's partial differential
  equation (Lagrange's PDE), 7, 18
Lagrange–Charpit method, 17, 32
Lagrangian, 42
Laguerre polynomials, 84, 91
Laplace equation, 54, 58, 65, 114
Laplace(1D), 99
Laplace(2D), 100, 161, 185, 192
Laplace(3D), 102
Laplace(circular), 127
Laplace(cylindrical), 86, 117

Laplace(rectangular), 115, 116
Laplace(spherical), 118, 131
Laplace transform, 151, 167
Laplace–Mellin integral formula, 151
Laplacian, 58, 67
Legendre functions, 210
Legendre polynomials, 84, 89, 90, 211
Legendre transformation, 42, 44
linear, 3, 7, 51, 144
Lorentzian, 155

Mellin integral formula, 151
Mellin transform, 151, 176
metric, 66
mirror image, 129, 132
mixed-boundary condition, 192
modified Bessel functions, 114, 208
modified Helmholtz equation, 114,
    191
Monge cone, 16–18

Neumann condition, 81
Neumann functions, 92, 203
normal vector, 9, 10, 12, 14, 15, 60, 95

oblate spheroidal coordinates, 73
order, 3
ordinary differential equation (ODE),
    2, 4, 8, 10, 66, 71, 80, 84, 98, 172
orthogonal curvilinear coordinates, 67
orthogonality, 81, 89–91, 205, 211,
    212
orthonormal polynomials, 89

parabolic cylinder coordinates, 78
parabolic partial differential equation
    (parabolic PDE), 59
paraboloidal coordinates, 75
Parseval's relation, 93, 159, 179, 190
partial differential equation (PDE), 1,
    51
particular integral, 5, 24
particular solution, 5, 24
periodic boundary condition, 86
periodic function, 169
plane wave, 11, 138

Poisson equation, 54, 58, 97, 116
Poisson integral formula, 130
principal solution, 98, 106, 110
prolate spheroidal coordinates, 71

quasi-linear, 7, 12, 18

rank, 3
reciprocity principle, 106, 123
rectangular (Cartesian) coordinates,
    66, 67
rectangular membrane, 147
recurrence formulae, 206
region of dependence, 65
region of determination, 65
region of influence, 65, 108, 120
Rodrigues' formula, 210
rotation (rot), 67

self-adjoint differential operator, 96
semi-infinite, 125, 161, 174
separable condition, 68
separation of variables, 28, 35, 66
singular integral, 24
singular solution, 24
solution, 8
spherical Bessel functions, 133, 207
spherical coordinates, 74
spherical surface harmonic functions,
    209, 213
spheroid, 2
stationary point, 37
Stokes' method, 164
Sturm–Liouville equation, 81, 84

tangent vector, 13, 14, 60
toroidal coordinates, 80
transient phenomena, 173

unit step function, 95, 108, 154

variation, 38, 39
Volterra equation, 194, 195, 197

wave equation, 52, 63, 65, 114
wave(1D), 57, 87, 106, 118, 146, 164

wave(2D), 57, 147   wave scattering, 138
wave(3D), 57, 108, 120
wave(circular), 133   zero points, 82, 92, 134, 136, 205

Printed in the United States
by Baker & Taylor Publisher Services